Environmental Forensics and its Effects on Investigations

Environmental Forensics and its Effects on Investigations

by

Greg MacMaster

Cover graphics by: zeegs80@gmail.com

Copyright © 2006

All rights reserved. Printed in the United States of America. No part of this publication may be reproduced, stored in a retrieval system, or transmitted, in any form or by any means electronic, mechanical, photocopying, recording, or otherwise, without the prior written permission of the author.

ISBN 1-58961-287-6
Published by PageFree Publishing, Inc.
109 South Farmer Street
Otsego, Michigan 49078
1-866-GO-BOOKS
www.pagefreepublishing.com

Contents

Dedication 7

Purpose Forensic Meteorology..9

Chapter 1: Want To Be A Forensic Meteorologist?....................13
Chapter 2: Accessibility & Availability Of Weather Data................31
Chapter 3: Legal Cases, Terms and Concepts..........................47
Chapter 4: Pathology..87
Chapter 5: Forensic Entomology Environmental
   Impacts in Estimating time of Death........................113
Chapter 6: Forensic Toxicology 143
Chapter 7: National Guidelines for Death Investigation.............149
Chapter 8: Aviation Accident Investigations...........................189
Chapter 9: Engineering / Construction 227
Chapter 10: Automotive Accidents and Impacts from
   Weather Disasters..243
Chapter 11: Sample cases from the field of forensic &
   meteorology 267
Chapter 12: Resources, Websites, Computer Programs &
   Spreadsheets...275

Bibliography...279

Book Reviews..283

Dedication

I want to thank a number of colleagues and associates that have shared information through the years keeping in line with the professional standards that we strive to maintain. Chief Medical Examiner Investigator Matt Kempf, who planted the seed; Dr. Matt Houghton—your support and backing helped keep me going and assisting in my collection of data. Jim VanAllen, my associate, buddy, and cousin—for if it was not for his ability to put my thoughts into a computer program and perfect it, part of this book probably wouldn't have made it off the ground. My wife, Denise, her tireless effort in keeping me focused on what's most important in my life, my family. Thank you!

Purpose
Forensic Meteorology

"What is a forensic meteorologist" is something I get asked quite often and the image of the broadcast meteorologist on TV is what most people envision. After an explanation of what a forensic meteorologist is and what they do, they think CSI the hit TV series, and to a degree I can relate our profession to some shows where the weather played a major factor in deciding someone's fate, although not as glamorous. There are clues in every case and weather is one of them. It can help narrow down a time. It can be a cause of an accident and it can even rule out a prime suspect.

Forensic meteorology is the scientific application of meteorological analysis to cases involving events for which legal and/or insurance claims are being made. The forensic meteorologist "reconstructs" the weather near the time and place of the event, using and interpreting all available data sources, including weather observations, weather radar, and satellite, lightning detection equipment and any other information that is deemed necessary to the case.

All available data sources can and should include **non-traditional** methods of weather recording from armchair weather enthusiasts, local farmers, schools, department of transportation records or anything that can lend a factor in the analysis. The elements of weather have a dramatic influence not only on our future but also on our past.

There is no "distinct" definition of "forensic meteorology" that is analogous among publications and opinions widely vary.

In its simplest form, forensic meteorology pertains to the law or legal system.

Some colleagues I conferred with agreed. So what follows is what was presented at the American Meteorological Society Forensic Workshop, San Diego, CA. 2005 by Sean Potter, CCM.

Forensic meteorology as defined by the World English Dictionary, (North American Edition)

Forensic: From the Latin word forensic, meaning "of the forum" relating to the application of science to decide questions arising from crime or litigation.

Forensic Meteorology: The application of meteorological and climatological data to legal cases and related investigations in which weather may have been a factor.

The role of the forensic meteorologist: "To assist the court in the *search for the truth* by presenting the *most accurate* description possible of the meteorological events pertinent to the case of litigation." (Haggard—2003)

"To fulfill this role, the forensic meteorologist must utilize the *best possible data set* in his or her retrospective reconstruction of the weather. The quality of the analysis is dependent on the availability of *representative measurements* of meteorological parameters." (Haggard—2003)

(As presented by Sean Potter, CCM, at the AMS Conference Workshop for Forensic Meteorologist, San Diego 2005)

The forensic meteorologist, who may act as either a background consultant or a testifying expert, will collect, interpret and analyze atmospheric data in support of insurance fraud claim investigations, civil and criminal trials, and environmental regulatory actions. The forensic meteorologist may be employed directly by an insurance company, the attorneys for either the plaintiff or defendant in a case or, with increasingly frequency, may be appointed by the court itself. Regardless of the employing party, it is not the role of the meteorologist to be an advocate for either side in a dispute, but to assist the judge and/or jury in understanding the often complex facts in a case so that they may reach an appropriate verdict.

The legal definition of **Expert Witness**: When knowledge of a technical subject matter might be helpful to a trier of fact, a person having special training or experience in that technical field, one who is called an expert witness, is permitted to state his or her opinion concerning those technical matters even though he or she was not present at the event.

Some typical problems we (expert witnesses) deal with in forensic meteorology are pretty typical and range from the following:

"The automobile accident was caused by poor visibility—was that caused by natural fog or pollutants from a nearby industrial plant?"

"Was the building damaged by a tornado or straight line thunderstorm winds?"

"A person was found electrocuted near a downed power line—was it a fault in the utilities' line or a lightning strike?"

"How can we demonstrate that rain fell at a site that is located many miles distant from any national weather service reporting station?"

The forensic meteorologist may collect standard and non-standard weather observations, assemble weather radar and satellite imagery, process weather data taken by a party in the case, or locate nonstandard sources of data such as lightning ground strike reports or atmospheric data taken by air pollution monitoring networks. This accumulated data is then used in a comprehensive analysis of the meteorological facts pertinent to the case. There is increasing use of sophisticated computer graphics and video animation of weather information in trials and mediation/administrative hearings. Most forensic meteorologists have had long and varied careers in the atmospheric sciences, and it is their hard-earned expertise that is in demand. Most successful forensic meteorologists have met the qualifications of a Certified Consulting Meteorologist (CCM).

Now that we have a fairly good idea on the definition and role of the forensic meteorologist, how does it affect other forensic scientists in their field of study? How much of an influence does our research have on the entomologist (study of insects), engineer, construction company (residential and commercial), transportation (aviation, automotive, rail and marine), pathologist (medical examiner) and law enforcement?

This book is a first attempt to introduce the ancient art of forensic meteorology, its impact and influence on other forensic sciences.

"Environmental Forensics (Forensic Meteorology) and its effects on Criminalistics" is an excellent reference for other forensic professionals who are looking to improve their investigating technique in cases where the weather would be a factor. While some information may not be directly related to a meteorologist, it does serve as a tool to understand what scientists are faced with when conducting investigations and gives the meteorologist a keener sense of awareness when conducting their own investigation.

Chapter 1

WANT TO BE A FORENSIC METEOROLOGIST?

It was mid-afternoon Monday (3 P.M.), cool and dry with a mix of sun and clouds when the groundskeeper heard two dogs, chained up and barking uncontrollably towards the house. He walked over to investigate why the dogs were barking and, peeking through the glass, noticed a body lying on the floor face down. He dialed 911 from his cell phone and police and emergency personnel soon arrived to determine that due to the advanced rigor mortis state, the victim had been dead for quite some time. They radioed for the medical examiner who, after an hour, officially pronounced him dead at (4:30 P.M.). He then took a body core temperature. It was 86 degrees. The room temperature was reported to be 75 degrees, however, the electronic thermometer changes at 11 p.m. to 70 degrees. He was found on the main floor, there was also a basement and second floor. The house was immaculate, clean to the touch with no dust anywhere. It was obvious he was a neat person. The house was heated with baseboard heat and all windows were closed and locked, except for the door to the backyard. As a precaution, the ME took an outside temperature of 62 degrees at 4:45 P.M. Light winds and partly sunny.

The extremities of the deceased were fixed, hard to move. He was wearing light clothes and for the exception of a small cut on his face which was revealed after they rolled him over, inspection indicated blow fly eggs had been deposited. After a cursory inspection of the grounds around the house, they found nothing highly unusual except for the open shed door in the back with a few items out of place. They did notice some marks which indicated the dogs dragged a large stick, which was close by, across the yard.

In the course of the investigation, the local meteorologist gets a call and is asked to provide the local weather conditions for that day in question. He obliges and gave the high and low temperature for the day, .16" of rainfall and that winds were gusty in the early morning hours. But was that enough for the investigators to make a determination on estimated time of death? Was it a natural death, homicide or murder?

That's not for us, as forensic meteorologist, to decide. But having knowledge in other facets in forensics would give clues as to what may have happened and in the process, give us a better idea as to what to provide the investigators. The deceased was a neat person, living clean, yet fly eggs were found in a facial cut that was against the floor? Drag marks in the yard suggesting the dogs may have played around (yet they were chained up). Questions start to surface to whether he was killed outside, and then moved indoors to hide the crime. He may have been outside long enough to attract flies and then was moved indoors. That would explain the fly eggs and drag marks. What about body temperature? The body cools at an average of 1 to 1.5 degrees per hour.

If you had a working knowledge in how a body cools and estimating time of death, you would find a conflict in some findings.

1. The body temperature indicates that he would have been dead over 24 hours (if compared to the temperature inside the house).
2. Using outside air temperature (62 degrees) in the formula, you find that the estimated time of death, (ETOD) would put it at 13.2 hours to 18.8 hours (between 9:42 P.M. the night before to 3:18 A.M. the following morning). But as a forensic meteorologist you notice something in greater detail that could decrease the timing error.
3. Meteorologists are aware of how winds make us feel cooler and know that water cools the body three times faster. Formulas can prove this theory. So what do you do?

I did a study in my own house on temperature and here's what I found;

Main floor thermostat is located on main floor (5 feet high)

Second story bed	70.7
Main floor wall	69.7
Placed on carpet /main floor	67.9
Basement floor	65.7

There's a 5 degree difference between the basement floor and the second story bed. Would this 5 degree difference change the ETOD? If inside temperatures vary this much, how much of an impact would a change in outside temperatures have?

Education, skill, working knowledge and experience, by far, are the most important factor in determining if you want to be an expert witness. Simply researching archived weather data isn't enough to label you a forensic meteorologist (FM). It's how that data is applied to the research and findings. To search the past correctly, you have to know how to forecast.

Knowledge in forecast models, upper air soundings, surface observations and other areas help you understand the fluid dynamics of how the atmosphere works. This will heighten your visual perception when you appear at a crime scene and have to find topographical influences that may have been overlooked by the accused, their legal representation or other professionals who may have been called as "expert witnesses."

With that said, you should have a minimum of educational requirements to enter into the forensic meteorologist profession. A degree in the atmospheric science field, preferably meteorology is preferred. Other fields of discipline within the atmospheric science field may be:

Applied climatology
Meteorology as it applies to aviation, marine
Mountain wave
Solar
Oceanography
Satellite meteorology
Radar meteorology

Thermodynamics and physics
Applied meteorology
Instrumentation
Dynamics

It would be a good idea to have an understanding of atmospheric processes on all scales. Above all, know your professional and personal limitations. Lack of knowledge can hurt the case, your credibility and your future. Additionally, courses in business (marketing, accounting and business ethics) as well as legal courses would strengthen your knowledge in your forensic business. If you want to expand your knowledge in other areas, such as, medical, building construction, law enforcement, aviation and so on, take classes and work in that environment and see how weather plays into that science.

You can never get enough education. Most government and private sector employers offer in-house training. Seasonal workshops are becoming more prevalent in the National Weather Service and often use analysis techniques to determine what caused a/an excessive rainfall/snowfall, severe weather event. This type of training is invaluable! There are a number of web and computer-based courses to help you gain knowledge in areas where you feel you may be lacking in knowledge. Scientific literature in your specific field of study will enhance your technical knowledge. The American Meteorological Society is the foremost authority in education criteria with established scientific manuals designed to assist you in this profession.

Experience: How does one go about getting experience? Well, don't expect to be a forensic meteorologist right out of school. It takes years of on-the-job training to develop skills necessary in the research field. You learn by your verified forecasts and busts, more commonly termed (successes and failures.)

A way to gain experience:

1. Professional experience in applied research and consulting,
2. Teaching experience—You will be called upon to instruct legal counsel, judge, jury or other professionals on information

pertaining to your findings. This is not the time to be nervous, so you'll need to be comfortable
3. Private, governmental and corporate world
4. Years of (non-forensic) work experience
5. Working under an expert witness helps shape the mindset in forensics
6. Review your forecast and research conditions that changed and led to a missed forecast.

Experience also helps you expand your knowledge in how the weather impacts our daily lives. Simply forecasting isn't enough, but how that forecast impacts our pattern of living day to day, week, weekends, shopping habits, travel itinerary, vacations, farming, harvesting. I could go on forever. This is how you develop a better appreciation how sensitive we are to the weather. This sensitivity helps us to find the truth in research.

Other ways you can educate yourself on other professional fronts— enroll in classes designed for that specific profession at a local college or University. In the EMS/Fire field, classes are held annually to maintain their credentials. There are many ways to increase your knowledge in other professions. But it takes drive and ambition.

So what do we do as forensic meteorologists? (See Chapter 3 on the specific steps of protocol). Well, to start, we acquire data and interpret it for those who are unfamiliar to the "weather lingo." We have to simplify the terminology so it's easily understandable in a court of law. Keep in mind that you may have to present your finding in front of a jury. These people are from all walks of life, ethnicity, income and professional status. Your jury may consist of a baby-sitter/nanny, heart surgeon, boat captain, college student, secretary, retired, school teacher . . . you get the idea. This is where your ability to teach/present comes into play. You also will advise clients on the importance of the weather and how it may have played a factor in the case. You may have to perform a multitude of specialized analyses in order to find a variable to assist in your findings. (Upper air dynamics, thermodynamics, micro-analysis, etc)

One of my more popular cases which defined my credibility in forensic meteorology came with a call from a lawyer asking for the wind data on a specific day. This data came from a local airport and the site of the house fire was miles away (as the crow flies). The airport, which lies in a valley oriented north—south, showed a wind out of the south at seven to nine miles per hour. When I learned more about the case and the questions posed by the attorney, I found that I would have to do a micro analysis from the upper atmosphere to determine the wind flow at the surface. The regional wind plots showed varying directions and speeds and the accident (two houses caught fire) occurred on the east shore of Lake Michigan.

To make a long story short, had I used the airport data (single source); one insurance company would have saved millions of dollars in payout. Instead, the data and supporting evidence I had to substantiate my analysis, along with formulas and operational techniques, allowed me to come to the conclusion that the wind flow was from the west. This shifted the responsibility to the other insurance company for payout. But the legal ramifications could have been much worse. Had I presented the wrong conclusion, fault could have fallen on the builder. The future of a family business was hanging on the testimony.

Evaluating a report from the opposing weather expert witness (or other witnesses) may shed light on a situation which could help strengthen your case. I have found it best to prepare my expert witness draft report before I review the opposing legal testimony. Review with your legal counsel and then finalize your draft for submission. At the end of this book I have a sample paragraph which will leave options open to submit additional information at a later date.

As you can see, experience comes from years of working in the field and being subjected to the daily routine of your profession. Don't expect to be a success or get rich overnight. It's a great second income and should you decide to venture into a full-time business, expect slowdown times. In other words, have another source of income until you become well established.

There are thousands of meteorologists nationwide employed by corporations, governmental sectors, broadcast stations and self-employed. How do you know if you have the qualities to be a good forensic meteorologist? Well, for starters, you must have the desire to find the truth through tedious hours of research for little or no pay. If you're getting into the business for money, you're getting in for the wrong reasons. If you're a big fan of CSI, Forensics and Legal / Court Files, then you have a running start.

If you would like to test your skills and see if you have the fortitude, try interning with a forensic meteorologist. Accompany him/her to accident sites and ask questions (but don't get in the way). Review previous cases with them.

Here's a site where you can share your thoughts/questions or if you're just looking for some insight.

http://groups.yahoo.com/group/forensic_meteorology/

Finally, there's respect for our fellow forensic meteorologists, as it's not uncommon to have an expert witness on both sides (plaintiff and defendant). We can't share ideas and thoughts so we work independently and for good reason (conflict of interest). Keep comments about the opposing testimony professional and respectful. Remember, we all have the right to make modifications and changes should we become aware of additional information that could sway our testimony. If you're unfamiliar with an aspect that you feel would help your testimony, seek the advice of a fellow forensic meteorologist (not associated with the case) who may have better insight to that specific science, (air pollution and chemical dispersion) as an example. Be prepared to offer up a portion of your pay to offset the cost for information. Remember, they're in the business to make money, too.

Private-Sector Meteorological Consulting: View through the Lens of a Sole Practitioner

Phillip D. Falconer, C.C.M.
Falconer Weather Information Service
Scotia, New York

Invited Presentation
"Vision Seminar Series"
National Weather Service Headquarters
Silver Springs, Maryland
November 8, 2000

I am gratified that Gen. Kelly and the organizers of the National Weather Service's Vision Seminar Series have invited me to appear before you, to represent the views of another of your constituents from the ranks of the private sector, namely that of the meteorological consultant. As much as I might wish to consider myself a visionary, I feel best qualified to share with you a few thoughts I have on the present-day health of the weather consulting business, and touch on a few challenges and opportunities I see for this segment of the private sector in the coming few years.

I know that you have previously heard from representatives of some of the country's largest and most prestigious operational forecasting firms and value-added weather information suppliers; their views on the appropriate roles of the National Weather Service and of the commercial weather industry are important to hear and discuss. However, the business of weather in the United States is diverse, and the views of the commercial weather industry are not necessarily joined by each of the various constituent groups comprising the private sector meteorological community. Indeed, no one segment of the private sector can lay claim to a unified view of the appropriate roles and responsibilities of the National Weather Service and its private sector partners.

For many meteorological consultants, most of whom are not directly engaged in operational forecasting, recent discussions regarding future role of the National Weather Service in the provision of our country's weather and climate services, while evocative, have seemed at times both remote and unsettling. With interest in the topic running high, it was not altogether surprising that in July, 1999, the American Meteorological Society adopted a Policy Statement on the Public/Private Sector Partnership, expressing its view on the appropriate roles and shared responsibilities of the government and of the private sector in the provision of this Nation's weather, climate, hydrologic and environmental services. This policy statement is, in my opinion, a balanced and reasoned view of the core issues surrounding this partnership, and was developed in conjunction with the nine-member AMS Board of Private Sector Meteorologists.

The Business of Weather Consulting Comes of Age

Recent estimates, although rather uncertain, suggest that there are perhaps 300 to 400 'for-profit' firms nationwide engaged in the business of weather, ranging in size from individual consultants to corporations employing dozens of meteorologists and enabling support technologists.

The private sector meteorological community is engaged in a stunning array of service initiatives, including, but certainly not limited to, the sale and delivery of basic weather and climate data; the creation of value-added meteorological data products and enhancement of weather data visualization capabilities, operational weather forecasting, meteorological and environmental consulting services, and the production of weather monitoring, communications and computing technologies. Many private meteorologists work directly for weather-sensitive industries, including the gas and electric utilities sectors, the airline, transportation, shipping, and agribusiness communities, the broadcast media, and in commodity exchanges.

While there is a rich history surrounding the humble beginnings and gradual development of the commercial/industrial weather sector over the past half century, the pace of growth in the private sector, and the applied technological innovations it has contributed to the science of meteorology, has noticeably accelerated during the past decade. As we enter the 21st century, roughly one out of every three meteorologists has found employment in the private sector. By the year 2005, the number is expected to swell to 40 percent. And, by the end of this decade, nearly half the nation's meteorological work force will likely be privately employed. In the competitive marketplace, this growth will undoubtedly spur a host of changes in the manner, efficiency and costs of delivering applied, customer-specific weather information to the nation.

'Boot Camp' for the Forensic Meteorological Consultant

Much of the meteorological consulting work I have done since 1980 or so involves the acquisition and analysis of past weather records for insurance claims investigations, and matters of tort litigation. Those who have been involved in the forensic meteorology specialization have seen this segment of the profession grow steadily over the past twenty years. Several dozen firms, many of which operate as sole proprietorships, now offer forensic meteorological services; some are able to do this profitably on a full-time basis.

Other than three American Meteorological Society workshops on the application of weather information in legal proceedings, held between 1989 and 1996, and a few generalized primers on the presentation of scientific evidence in court, the meteorologist is hard-pressed to find any opportunities to learn about the intricacies of the U.S. judicial system and the manner in which expert weather witnesses are expected to participate in this system. This points out, of course, a need which is likely to be filled in the next year or two. Presently under discussion at the National Council of Industrial Meteorologists is how we might facilitate one or more workshops, where meteorologists would be able to fine tune their courtroom skills in a mock trial setting.

Participants would be introduced to a hypothetical law suit involving some aspect of the weather; would watch the matter litigated; and would then have the opportunity of evaluating the proceedings. Trials would be held in a mock courtroom, with a presiding trial judge, opposing attorneys and meteorological experts, and demonstrative exhibits. Such a continuing educational program would touch on each of the key elements of forensic weather investigations:

- pre-trial depositions and expert affidavits;

- courtroom rules and procedures affecting expert witness testimony;
- the introduction and use of meteorological evidence at trial;
- the objectives of direct examination and cross-examination of the weather expert; and
- the role of the weather expert as an educator in the courtroom.

There has been talk about facilitating such seminar opportunities for a few years now, especially in view of the fact that previous workshops were largely a pot pourri of meteorological case studies and 'war stories', often with limited discussion of the key legal aspects of the case; of the lawyer-expert deliberations prior to and during trial; the manner in which demonstrative meteorological evidence was used, or the scope and general substance of opposing counsel's cross-examination tactics. Any one of these topics would provide useful guidance to those generally unfamiliar with the 'drama' of litigation.

The time has come for our profession to encourage 'hands-on' training opportunities for those meteorologists who find themselves increasingly involved with weather-related insurance claims investigations and expert testimony. This means collaborating with representatives from the claims industry and legal profession, not only to insure that meteorologists understand what is expected of expert witnesses, but for the insurance industry and attorneys:

- to understand how, when and where weather information is collected;
- how it is disseminated to the public;
- what the various weather data products mean; and
- why the selected records bear particular relevance to the case being litigated.

Harvesting the Nation's Weather Data Resources: Positive Steps-Ongoing Challenges

Continuing this line of thought, I wanted to touch on an issue that bears obvious impact on the meteorological consultant's ability to document past weather conditions, based on the present-day ensemble of NWS data products. As a result of the agency's multi-year modernization program and as generally set forth in its recent Strategic Plan for Weather, Water and Climate Services through 2005, weather service customers and partners now enjoy convenient and direct internet access to a wide array of operational data sets, charts, images, and technical reports. Not too long ago, many of these products were available, by request, only in hardcopy format through the archives of the National Climate Data Center (NCDC) or through third-party commercial weather information service providers; this proved to be a costly, often time-consuming expense for meteorological consultants, particularly in situations where their clients required expedited service. More open access to a wider variety of Weather Service documents and products has allowed many meteorological consultants to build their own digital libraries of weather, water, and climate information, thereby enabling them to more efficiently pass along operational data and value-added products to their customers.

However, with as many benefits as have arisen out of the NWS modernization efforts, the reality for many meteorological consultants is that today's easy access to more National Weather Service data products has been offset by some operational and reporting limitations in both the national network of automated Surface Weather Observations systems and in the Hourly Precipitation Data station network.

Over the course of the past five years, those of us who routinely use surface weather observations in reconstructing past weather events are painfully aware that, while there are now more reporting stations across the country, some providing more frequent observations than ever before, the loss of, for example, the six-hourly synoptic reports at the automated surface observing system (ASOS) stations, particularly snowfall and snow depths; the lack of thunderstorm reports at many stations; and the replacement of nearly all of the universal weighing rain-gauges with (lower-resolution) Fisher-Porter-type gauges across the nation's rain monitoring network have had the unfortunate and unintended consequence of jeopardizing the weather analyst's ability to fully-evaluate past weather conditions.

It is not my intention to go into a litany of concerns that have been expressed over the operating characteristics of automated weather observing systems, or of the Fisher-Porter rain gauge. But institutional changes in this country's basic weather observing infrastructure has, and will likely continue to affect those of us who rely so heavily on National Weather Service data products, particularly in the 'snow and ice' belts of the United States, where we often have to make do using proxy indicators and supplemental private sources of weather information.

This does not go unnoticed by our clients, in part because the conscientious weather consultant may need to advise them of any known limitations in the basic National Weather Service measurements or reporting procedures, which might compromise an expert reconstruction of past weather conditions. Again, I mention these matters not so much as a criticism of our present-day national weather observing infrastructure, but as an opportunity to provide you with feedback from a private sector constituent.

Desktop Computing Power Unleashes New Opportunities for Weather Consultants

The arrival of fast, affordable, powerful desktop computers; of wide arrays of digital data sets; and of gigabyte-sized data storage devices has enabled meteorological researchers and modelers to quickly retrieve and process the large volumes of data needed to run specialized weather forecast, atmospheric flow simulation, and air pollution transport and dispersion models. These technology developments have also allowed meteorologists to archive vast amounts of historical weather data and data images in digital format, which may then be easily retrieved for application to specific case studies, or to broader climatological or statistical analyses. With these tools, the consultant is able to offer customers a broader and more sophisticated selection of work products than were ever before possible, often on a faster turn-around schedule.

The data ingest and processing speeds necessary to view animations of Doppler weather radar, high resolution satellite images, and lightning ground strikes has been well within the capabilities of desktop computers for many years. More recently, sophisticated, computationally-intensive predictive and diagnostic models of the atmosphere have been developed, which now can run on stand-alone PCs. One Minnesota-based commercial firm uses a PC-based version of the NCAR-Penn State University MM5 mesoscale meteorological modeling system, to generate real-time weather forecasts as well as three-dimensional flow visualization and pollutant transport simulations in complex terrain. Another forecast consultant now provides sailors and outdoor adventurers with customized wind and weather forecasts, based on the ETA model, which he runs on a Linux PC workstation. A Georgia-based company markets a Windows-based hurricane tracking, analysis, and planning software program, with numerous capabilities tailored for the emergency management community, including SLOSH storm surge data; a SLOSH inundation level analysis subroutine; and various user-selectable impact assessment tools.

Soon analysts will be able to routinely explore geographically-referenced environmental data sets; to test various 'what if' storm threat scenarios; and to search for and evaluate spatial relationships between meteorological and environmental properties using complex queries. Work is underway within NOAA's National Data Centers, especially development of the NOAA Virtual Data System, which employs new GIS technologies and GIS-enabled databases, will facilitate online, browser-based information mining.

It is not only likely that opportunities will arise for the private sector meteorological community to partner with government and academic institutions in developing the visualization tools and underlying geo-referenced environmental data sets, but that commercial weather enterprises will also enter this market on their own, capitalizing their risk entirely with monies derived from the private marketplace.

Minding Our Business in the Classroom: A Plea to Develop University-Level Opportunities Relevant to the Commercial Weather Service Industries

I mentioned earlier there should be ongoing opportunities for meteorologists working in the forensics area to improve and sharpen their skills, by participating in focused workshop practica designed around the concept of the 'mock trial.' But there is fundamentally a much broader need for specialized education related to the private practice of meteorology. Undergraduate atmospheric sciences curricula at universities across the nation simply have not done a particularly effective job in teaching students about the business aspects of applied meteorology. Many meteorology students today will, upon graduation, work in the private sector; it is disappointing to report that most will not have had the opportunity to learn the fundamentals of sound business practice, of customer service, or even how weather affects industry operations and its infrastructure.

While unsettling, this is not particularly surprising, as relatively few atmospheric science faculty have first-hand, private sector experience in operational meteorology or with the business aspects of full-time meteorological consulting. The University of Oklahoma at Norman has tackled this issue head on. Recently, the School of Meteorology and College of Business have collaborated to offer an undergraduate area of concentration in private sector meteorology. WeatherData, Inc. of Wichita, Kansas has also attempted to address the issue, by organizing a 3-1/2 day workshop, "Introduction to Commercial Meteorology for University Faculty" designed to illustrate how education centered on real world, workplace-oriented challenges is important to success in the private weather industry. Faculty workshop participants partner with company meteorologists, who review how weather affects their clients, and guide participants through the process of preparing client-specific weather forecasts on short deadlines, alerting their customers to pending severe weather, and handling research requests. This is precisely the nature and scope of effort needed in our universities to prepare the next generation of meteorologists not only for work in the commercial weather industry, but also across the spectrum of private sector meteorological opportunities, including the business of weather consulting.

Some Thoughts on National Weather Service Participation in the AMS Certified Consulting Meteorologist Program

Akin to the licensing of doctors, attorneys, accountants, engineers and other professionals, the Certified Consulting Meteorologist (CCM) program of the American Meteorology Society assures the public that the individual possessing the CCM designation has successfully completed a rigorous peer-testing evaluation, and has met or exceeded certain high standards of technical competence, character and experience. Since its inception in 1957, the AMS has issued over 600 certificates, most of these granted to private sector consultants, industrial meteorologists, and atmospheric scientists from the university community and, to a lesser extent, to meteorologists employed in state and federal agencies and government-supported research institutions.

Relatively few National Weather Service meteorologists have ever applied for the CCM designation. Why this is so is unknown, but may relate to any one or more of the following reasons:

- confusion over the meaning of the term 'consulting meteorologist' and whether certification as a consultant would be perceived to be in conflict with, or otherwise jeopardize the government meteorologists' traditional public service function;
- the fact that the NWS already administers several internal continuing education and training programs leading to professional qualification;
- that relatively few NWS field staff are members of the American Meteorological Society; or
- that the Society's marketing efforts for the CCM program have simply not been effective.

Whatever the reason, there continues to be few National Weather Service certificate holders.

You may ask, then, why should I, as a private sector consultant, be at all interested in raising the visibility of this particular professional certification program within the Weather Service? I believe the CCM designation represents a fair and reasonable opportunity for those in both the private and public meteorological communities, those who in one fashion or another 'consult' with their constituents, to demonstrate a common minimum standard of technical excellence; parity, if you will.

The current AMS Certification Board includes representation from the National Weather Service, and efforts to attract operational meteorologists are under active consideration. Many CCMs have voiced their belief that the Board testing procedures could be tailored to a candidate's area(s) of technical competence, including operational forecasting and weather analysis. In my opinion, the pool of examination questions and case studies for operational meteorologists seeking the CCM designation, should be updated and expanded, possibly in coordination with the National Weather Service training facilities in Norman and Kansas City and with the COMET program office.

I believe that where the professional interests and opportunities for technical achievement in the private and public sector meteorological communities intersect, a healthy mutual understanding of each other's abilities would emerge. The CCM program may represent just such an opportunity. Whether it deserves a second look by National Weather Service meteorologists, who may not have previously considered its relevance to their time in service, or who may have felt it merely a program for private meteorologists, remains to be seen.

Final Thought

Permit me to leave you with a final thought, that perhaps consideration be given to a simple, low-keyed program through which National Weather Service meteorologists, particularly those in your field offices, could be given leave time to shadow the activities of their private sector colleagues in the weather consulting business. Observation at close range of the consultant's daily workload, from the intake of client requests, to the acquisition and analysis of National Weather Service data records, and eventually to the preparation of value-added meteorological data products and reports, would go a long way in fostering an understanding and appreciation of the meteorological consultant's role in the business of weather in this country.

Chapter 2

ACCESSIBILITY & AVAILABILITY OF WEATHER DATA

The information super highway has opened up a whole new area in accessing weather information. Hardcopy records are still available at many local libraries, colleges and universities and some records are kept at the local level in governmental centers. This list is endless and it's up to you to be creative and think outside the envelope. Don't trap yourself with limited data. The more you have available, the better your decision making process.

First and foremost you will need official records, and to do this, you'll need access to the NCDC website. I would suggest you sign up for their newsletter and make sure they have your name for future enhancements and updates. It's a fee-based system, and charges vary depending on amount of data requested. You could use a more indirect means of access data for a preliminary review by going to *www.weatherunderground.com* and go to the history link. It's a quick view of what you could expect to encounter but shouldn't be relied upon as "official data." Some data may not be available, so try other searches for specific weather information. Some search engines used as of this writing (Google, Hotbot & Yahoo). Check with your local National Weather Service office and see what they keep on record and how long they keep it on file before they send it to NCDC for archiving.

LEGAL, INSURANCE, and CERTIFICATION
Information for legal and insurance claims
Information on data certifications.

What is "certified" data? Most states require all records that are to be submitted as evidence in a court of law to be authenticated in some way, typically called certification. There are several types of certification available. The National Climatic Data Center, the official United States archive for climatic records, can provide more comprehensive types of certification.

What can you testify to in court (and where can you testify)? Generally, it is not necessary for expert witnesses to appear in court to testify about the data we supply. If expert testimony is needed, the services of a forensic meteorologist should be retained. The American Meteorological Society and the National Weather Association maintain listings of consulting meteorologists.

Many outdoor events coordinators are looking to insurance companies for rain insurance. The threat of rain, in a measurable amount (usually .25" to 1.5) in a short period of time would insure a payout that would cover the loss of sales and/or damage to goods. Having a consulting meteorologist insures that the insurance company is getting accurate information that would otherwise be altered for financial gain. From outdoor concerts to antique auto shows, the amount of money to be lost from inaccurate reporting could be insurmountable to the insurance companies. So accurate reporting is a must.

An inspector said that there was hail damage to the roof of a home. "Can you tell me all of the dates that hail was reported in my area?" Well, that depends. First of all, hail is typically localized phenomenon. It may hail in one location, and a half mile away no hail may be observed. Because it can occur in a relatively small area, not all hail that falls is observed or reported. One of the criteria for a severe thunderstorm is observed hail ¾ inch or more in diameter, so if severe storms have occurred at your location there may be a record of hail if it occurred. These reports include the size of the hail, location where it occurred, and time it occurred.

For events less than three months old, check the preliminary daily storm reports on the Storm Prediction Center web. For events greater than three months old, check the on-line Storm Data database maintained by the National Climatic Data Center.

Official publications of storm data are compiled and published by NCDC. There is about a six-month lag in the publication of the monthly Storm Data publications. The minimum cost for receiving copies of Storm Data pages is $12.00 at the date of this publication.

If lightning data is requested, Global Atmospherics, Inc. is the place to go. They operate the National Lightning Detection Network, and provide custom reports on lightning strikes, and can locate strikes usually to within 500 meters. The NWS may be able to tell you whether or not a thunderstorm was reported in your area, depending on the availability nearby reporting stations.

What if you had wind damage to your roof during a recent thunderstorm, and the insurance company needs documentation of the wind speed before it will process the claim? Can you tell what the wind gusts were during the storm? Wind speed can vary greatly over short distances, and is affected by such things as trees, buildings, and other obstacles. Most wind speed and direction data is obtained from airport observations sites. Usually these have an unobstructed exposure to the wind and may not accurately represent the wind conditions at, for example, an urban location. If an airport observing station is located near your home, this data may be available and be sufficient evidence. How applicable this data may be to your location depends on your proximity to the observation site. If there is no wind observation site nearby, we would have to rely on the local storm reports gathered by the National Weather Service or the Storm Data publication published by NCDC. Local storm reports are listings of storm-related tornado, hail, and wind gust/wind damage events. Wind speeds reported are often estimates and usually not at "official" observing sites. However, these reports may provide evidence that a severe storm occurred in your area.

(Information or portions therein provided by H. Berg, Midwest Regional Climate Center)

At the end of this publication, you'll find links that can assist in your search for weather records.

Sometimes, data will be in raw form (numbers . . . and lots of them) and to decipher that data you'll need a computer program to crunch that data to give a more meaningful, simplified output. There's a plethora of programs to plot information and I'm sure I'll have to dedicate a chapter on this topic alone. We'll put this in the next revision of this book. I did a quick search and found a few (as of March/2005) which are referenced below:

http://grlevelx.com: **A complete radar display/analysis program, allows for viewing Level II (hi-res) and Level III (NIDS) radar data in real-time or from archives.**

www.raob.com: A powerful, multi-functional sounding analysis program that can read numerous types of rawinsonde/radiosonde data; create a variety of sounding diagrams; 3-D holograms; time & distance based vertical cross-sections; mountain (lee) wave turbulence; and display over 150 atmospheric parameters including icing, turbulence, wind shear, inversions—plus use of a unique severe weather analysis table, a cloud parameter table, and interactive diagrams.

www.wdtb.noaa.gov/resources/projects/BUFKIT/bufkit4.html BUFKIT is a forecast profile visualization and analysis tool kit developed by the staff at the National Weather Service (NWS) office in Buffalo and the Warning Decision Training Branch (WDTB) in Norman, OK. BUFKIT is used in the public, private, and educational meteorological sectors in both the US and Canada. BUFKIT is available to anyone who is interested in the analysis of forecast hourly profiles.

www.digitalatmosphere.com: Digital Atmosphere is powerful weather forecasting software that allows you to create detailed maps of real-time weather for anywhere in the world. It makes extensive use of techniques and algorithms that are comparable and in some cases superior to the National Weather Service's multimillion-dollar Advanced Weather Interactive Processing System. It can run on any 486 or Pentium system, and does not require any type of subscription or recurring costs, using free data from the Internet.

Archiving weather records is a cumbersome project and can take up tons of space on your computer and shelf space. So in lieu of the space problem, I have added a few links to assist in your research.

The U.S. Signal Office began publication of the maps as the War Department maps on Jan. 1, 1871. When the government transferred control of the weather service to the newly-created weather bureau in 1891 the title changed to the Department of Agriculture weather map. In 1913 the title became simply daily weather map. Eventually, in 1969, the weather bureau began publishing a weekly compilation of the daily maps with the title daily weather maps (weekly series).

This site offers maps from 2002 to the present.
http://www.hpc.ncep.noaa.gov/dailywxmap/frame.html

For maps dating from 2001 to 1871, go to this site: *http://docs.lib.noaa.gov/rescue/dwm/data_rescue_daily_weather_maps.html*

National Climatic Data Center
U.S. Department of Commerce

(The following information was referenced, in part from the NCDC website and the NWS)

NCDC is the world's largest active archive of weather data. NCDC produces numerous climate publications and responds to data requests from all over the world. NCDC operates the World Data Center for Meteorology which is co-located at NCDC in Asheville, North Carolina, and the World Data Center for Paleoclimatology which is located in Boulder, Colorado. *http://lwf.ncdc.noaa.gov/oa/ncdc.html*

You can link directly to an NCDC Weather Observation Station Record for a particular station (or list of stations) using the identifiers. Any call that finds more then one record will display a list of the records found. The user can then select the desired station from the list. To find only one record you should use call sign or WMO#. Follow this link to find out more:

http://www.ncdc.noaa.gov/oa/climate/linktowcs.html.

Certification information

The National Climatic Data Center (NCDC) is designated by Public Law 754, passed by the 81st Congress as the Federal Records Act of 1950, as the official United States archive for climatic data records. As an archive facility, the only fact that the NCDC can attest to is that exact duplicates of climatic records on file at this center have been provided to those that request such data. The standard Department of Commerce (DOC) or General certification of authenticity regularly provided to clients appropriately accomplishes this (see form examples). In accordance with 28 U.S.C. 1733, properly authenticated copies or transcripts of records or publications of the National Oceanic and Atmospheric Administration (NOAA) shall be admitted in evidence as equal to the originals thereof. Although this statute only applies to Federal Courts, many, if not all, States have similar provisions.

The DOC gold seal with blue ribbon certification of record authenticity has for many years been provided to thousands of clients who successfully entered documents as court exhibits in legal proceedings. This DOC certification has been developed by the NOAA Office of General Counsel, in conjunction with the NCDC and with the approval of each state's attorney general, to be universally acceptable to all U. S. federal and state courts. Only in extremely rare instances has this DOC certification been unacceptable to some courts, and then, there were only minor points of contention. In order to maintain an expeditious flow of data to requesters, certain individuals at the NCDC have been designated by DOC/NOAA to sign these certifications in lieu of the original signatories. These "for" signatures have also been accepted without qualification by federal and state courts.

The DOC certification, or any affidavit, provided by the NCDC in no way authenticates or guarantees the data values contained in the records provided. The NCDC cannot attest to data accuracy since it has no direct knowledge of observer expertise, instrument reliability, or the conditions under which the data were recorded. The DOC certification merely facilitates the admittance of the documents it is attached to as evidence in judicial proceedings. NOAA regulations prohibit employees of the NCDC from providing services to the public that would be in competition with private commercial enterprises. Prohibited services include consultation, interpretation, evaluation or any other services that are commonly provided by certified consulting meteorologists or other enterprises in the private sector. As data providers, NCDC employees can only advise customers of the availability and contents of data archived at the NCDC. Because of this prohibition, "expert witness" testimony cannot be provided by NCDC personnel. The NCDC cannot certify data obtained from any source other than directly from its archives, even if that data had previously been provided from those archives and returned to the NCDC for certification.

The **DOC certification** (Form CD-64) is a two-part certificate (certification and authentication) under the Seal of the United States Department of Commerce with blue ribbon. This is the highest form of certification offered by the NCDC and can be attached to a total of up to 40 pages of data. This type of certification is signed by both the NCDC Records Custodian and the NCDC Director as the *Certifying* Officer, *or their designated representatives.*

The **General certification** consists of an attached certification statement signed by a Certifying Officer and is acceptable in most courts. The statement is affixed with a staple to the assembled documents, up to 70 pages. On rare occasions, customers request certifications with wording that differ from that of NCDC's standard certification. This service is available but is considerably more expensive and it increases the time required to respond to the request to several weeks. The specific wording must be edited to correspond to NCDC's guidelines and restrictions, returned to the requestor for review, forwarded to the NOAA Office of General Counsel in Silver Spring, MD, for approval, and returned to NCDC for signatures. One (1) certification can be attached to up to 40 pages of data.

While this certification information is designed to respond to the most frequently asked questions concerning **weather** records, it is recognized that it is not all inclusive. Special circumstances may occur that require specific answers to questions, in which case the NCDC should be contacted for further information or clarification.

Official **weather** observations are taken, either manually or automatically, at several thousand sites throughout the United States. The locations of these sites in any state may be obtained from any National **Weather** Service (NWS) office in that state or from the National Climatic Data Center's web page at http://www.ncdc.noaa.gov/ol/climate/stationlocator.html .

Most stations record a daily precipitation measurement. At many stations, the daily highest and lowest temperatures are also recorded. Some stations have automatic precipitation gauges which provide a continuous record of rainfall. At approximately 700 NWS Offices much more detailed data are recorded at hourly intervals, such as wind, temperature, humidity, type of **weather**, atmospheric pressure, visibility, and clouds. Similar detailed data are recorded at some airports by the Federal Aviation Administration, airlines, the military, and contract **weather** observers.

The NCDC files forecasts and warnings (beginning July 1983), upper air data, accounts of major storms, radar images, satellite pictures, **weather** maps (both surface and constant pressure), observations from ships, buoys, and aircraft, limited solar radiation data, a variety of summaries of **weather** conditions, and climatological publications from foreign countries. Much of the information for the U. S. is published in the *Climatological Data* and *Hourly Precipitation Data* for individual states/areas and in *Local Climatological Data* for individual stations. A brief description of the most requested NCDC data sets is provided as Appendix A at the end of this section.

The original, filmed, and digital records of data from all types of stations are on file, and may be ordered from the NCDC. Information regarding specific data, services, and costs can be obtained by contacting the NCDC by one of the methods given inside the front cover of this brochure. Meteorologists or meteorological technicians are currently available by telephone for assistance from 8:00 a.m. to 6:00 p.m. Eastern time Monday-Friday. The Department of Commerce, including the NCDC, is required by Public Law 100.685 (Title IV, Section 409) to assess fees, based on fair market value, to recover all monies expended by the U. S. Government to provide the data to requesters.

Copies of materials will be appropriately certified upon **specific request**. If certified copies or additional certification of publications that carry a printed certification are needed, the type of certification desired should be specified in the original request. Uncertified copies cannot be returned at a later date for certification. A new order involving costs for the requested certified data is required. Orders for certified data can be placed online but current restrictions prevent electronic certification for online delivery. Online orders for certified data are processed and certified offline. Delivery is usually within 5–7 business days by the U.S. Postal Service.

Regulations generally preclude NOAA personnel from rendering opinions in private **litigation**. It is therefore suggested that attorneys obtain the services of a private consulting meteorologist (PCM) who can give an expert opinion.

Although the sunrise/sunset times do appear in some surface observation data and monthly summaries available from the NCDC. The Department of the Navy, U. S. Naval Observatory, Massachusetts Ave & 24th Street, Washington, DC 20390 has a web site where astronomical data can be found (http://tycho.usno.navy.mil.) There are many private sector sources that can be contacted for data concerning the rising and setting of the sun and moon, twilight, moon phases, etc.

MOST UTILIZED OF THE NCDC'S PRODUCTS

Surface Weather Observations

These ground level observations are taken at specific, mostly airport, sites primarily for, but not limited to, use in the aviation field.

Meteorological conditions are observed at least hourly and present sky condition, visibility, **weather** and obstructions to vision, pressure, temperature, dew point, wind direction/speed, and pertinent remarks. Other elements, such as relative humidity, are available for most major airport sites. The contents and format of these observations may vary depending on the site type and method of observation (manual, automated, etc.) Observations at some sites may be for less than a twenty-four hour period. Daily forms are available for several hundred sites.

Climatological (COOP) Observations

These are daily observations obtained from a volunteer network of trained observers. In addition to daily precipitation, most sites record the temperature at observation time plus the maximum and minimum temperature for the full twenty-four hour period preceding the time of observation. Some cooperative observations also record daily river stages, evaporation, soil temperatures and wind direction/speed. Monthly forms are available for about eight thousand active sites.

Unedited Local Climatological Data

Summarizes surface **weather** observation and daily data for the entire month. The data is an unedited version of the first page of the Local Climatological Data.

Local Climatological Data

Monthly summary of temperature, relative humidity, precipitation, cloudiness, wind direction/speed, and degree days; also contains 3-hourly **weather** observations and an hourly summary of precipitation for most sites. Annual publications contain a summary of the past calendar year as well as historical averages and extremes. Monthly/annual publications are available for about 300 active sites.

Climatological Data

Monthly editions contain daily maximum and minimum temperatures and precipitation. Some sites provide daily snowfall, snow depth, evaporation and soil temperature data. Degree day data are contained in each monthly publication with the seasonal (July-June) heating degree day and snow data contained in the July edition. The annual edition contains monthly and annual averages of temperature, precipitation, temperature extremes, freeze data, soil temperatures, evaporation, and cooling degree days. Monthly/annual publications are available for each state or grouping of states.

Hourly Precipitation Data

Contains hourly precipitation amounts from recording rain gauges. Maximum precipitation is presented for nine (9) time periods from 15 minutes to 24 hours for selected stations. Monthly publications are available for each state or grouping of states.

Storm Data

A chronological listing, by states, of occurrences of storms and unusual **weather** phenomena. Information on storm paths, deaths, injuries, and property damage are presented. An "Outstanding storms of the month" section highlights severe **weather** events with photographs, illustrations, and narratives. An annual tornado, lightning, flash floods, and tropical cyclone summary usually appears in the December edition, but may appear in a later issue. Monthly publication includes all states.

Monthly Climatic Data for the World

Contains monthly means of temperature, pressure, precipitation, vapor pressure, and sunshine for about 2,000 surface data collection sites worldwide. Monthly mean upper air temperatures, dew point depressions, and wind velocities are presented for about 500 worldwide observing sites. Monthly publication.

NATIONAL CLIMATIC DATA CENTER
151 PATTON AVENUE ROOM 120
ASHEVILLE, NC 28801-5001
TELEPHONE : (828) 271-4800
TD for DEAF : (828) 271-4010
FACSIMILE : (828) 271-4876
E-MAIL :
ncdc.info@noaa.gov
WEBSITE : *http://www.ncdc.noaa.gov*

"A national resource for climatic information"
(Courtesy NCDC website and NWS)

Notwithstanding, there are other means of data retrieval which can yield better results due to the obscurity of the location in relation to distant reporting station(s). This could be achieved by visiting local neighbors who may keep daily records of local weather phenomena. Some homeowners are so weather savvy that they have installed a weather station which records weather directly to their computer. Check with local farmers, agricultural extension offices and Department of Transportation offices. Some educational facilities such as public schools often have some form of weather recording devices for educational purposes. Marine environs almost always have a wind report for boaters. This is where you'll find information which could support the distant reporting stations or show how the local topography can greatly influence local tertiary effects. Although they may not be "official," it does help in obtaining the solution.

If all else fails and you have limited data to work from, you may have to utilize your forecasting and analysis skills and use upper air charts to analyze the surface features. This is one of the more difficult ways to obtain data for court room use, but if your analysis is proven accurate through scientific reasoning and it's theoretically and operationally sound, then you have a good chance of proving your findings. Such was the case in the house burning described in the previous chapter. It was this technique where I had to improvise using microanalysis techniques to determine what the wind pattern was at the surface. With the growth of remote sensing, this is becoming less of a problem in this day and age.

Video: After I spent over 40+ hours on a case, pulling my hair out trying to justify why the weather occurred at a specific time, I was presented with evidence that was submitted an hour prior to the courts. The next door neighbor had recorded this big house fire which showed the glowing embers cross the two-track from the west! Where was this video two weeks earlier? Needless to say, I still got paid and the video was visual proof that my theory was right and further substantiated my testimony. Not all cases are like this but it was nice to have that video. So, when you investigate the location, ask around. See if anyone has a record either by pictures or video.

Which brings me to another point, a picture is worth a thousand words, right? Well, take a trip and check out the area. Look for visible signs where the weather could have enhanced or retarded the accident scene. These are signs you can't see without visiting the site. Take pictures of the location in a 360 degree panoramic view. Check out topographical maps from the internet or other source. You're looking for anything that can directly impact the weather in that area. Example: A car spun out of control and was hit by another car. There were injuries and lawsuits were filed. The driver of the first car went down a hill (15% downgrade) and over-corrected due to the icy/snowy conditions. Since they were at fault, the driver of the second car who was not at fault sued the driver of the first citing negligence. To make matters worse, the passenger of the first car was also pressing charges against her own friend! Police cited the first car for driving too fast for conditions.

When I concluded my inspection of the area, I found a tree line to the north which protected/shielded the winds, gusting to 30 mph from the northeast that morning. As the driver came over the hill, she encountered strong winds from the northeast because she left the protection of the tree line and experienced what we call wind-sheer. She over-corrected due to the strength of the wind and that's how the whole mess got started. It was snowy and the road had patchy ice according to police records. Case was dismissed.

Had I not investigated the area, I would have had no idea as to the terrain *www.topomap.com*, topography or the effect caused by the uneven terrain. I'll reiterate; always try to visit the site and if you can't, employ someone with video/camera capabilities.

A sample topographic map pulled from *www.topomap.com*

In search for the truth, we often employ other methods which could alter our own simplified findings. Confused? Don't be.

Here's an example:

> A slip and fall occurred in the Christmas Eve morning hours (9 a.m.). She claimed that she lost her footing in 8" of fresh fallen snow and that there was glaze ice below the snow. The days leading up to the accident showed that it was warm, mid 40's and rain had fallen the previous day. During the overnight hours, the passage of a cold front chilled the air with a temperature drop from 41 degrees to 24 degrees within 8 hours. The rain had stopped just prior to frontal passage (Fropa) and the winds increased in speeds in excess of 25 mph.
>
> After .04" of rain and falling temperatures to 24 degrees, our basic review would indicate that the moisture froze and the onset of snow would indeed support the victim's claim that she

slipped and fell. Besides the overabundance of medical claims, there was something else that didn't seem right. I reviewed the accident scene (three years after the date of the accident) and took pictures. Reviewed them and compared notes to the weather observations and re-created the weather for that day. Oddly enough (and what are the chances of this happening) I was within two days from the same day of the accident and we had warm temperatures as rain was forecast. Only this rainfall was expected to exceed ½" within 12 hours, followed by cooling temperatures and, of course, snow. What better way to test my theory than to see it first hand?

I recorded the weather and took pictures before, during and after the rain as well as the onset of cooling temperatures and snowfall. One thing eluded me and that was the drainage in the parking lot. With over ½" of rainfall, the pictures showed that, over time, the asphalt was damp to the touch, not wet. I researched the criteria on asphalt paving construction rules for the state and measured the parking lot and found that the lot had exceeded the minimum requirements for drainage by an additional 42%. With sufficient drainage and pictures to show for it, I was able to conclude that there was sufficient drainage before the onset of snow. This meant there was no "glaze" of ice as the victim indicated in her deposition.

One additional note: Having researched the asphalt cooling thermodynamic laws, I came across a Pave-Cool computer program which shows the cooling rates of asphalt based on air temperature and sub-surface temperature. By inputting the variables I had observed, the program calculated the cooling rate of the parking lot from 42F degrees to 32F degrees to be 4.5 hours. Whereas the air temperature drop from 42F to 32F was only 1.2 hours. This finding further supported the negligible risk for ice formation on the surface of the parking lot. One last comment on this Pave Cool program: the program was initially designed for asphalt cooling above 180F degrees. After I consulted with the engineer, (who also happened to be the computer programmer) he was able to modify the program to suit my specific needs in this case (sub-freezing). Think outside the box!

It goes without saying that if you search for the truth; you often have to rely on your abilities to acquire information outside of your immediate profession and scope of knowledge. In researching, you'll come across information that could have helped a previous or current case if circumstances were different. Catalog and save anything you find that will assist you in upcoming cases. It will save you time. I have all my research saved in folders (Marine, Auto, Aviation, Construction, Human . . . and so on). It'll be much easier to find and confer upon in the future.

Chapter 3

LEGAL CASES, TERMS AND CONCEPTS

Due to the visibility of the **broadcast meteorologist** within the community, there's a really good chance you'll get a call or two on weather information which could eventually go to trial. To relieve yourself of any undo hardships, it would be best to review your station policy or handbook, let the station manager and/or news director know of your work and to insure that there would be no conflict of interest. They would be the deciding party on whether you take the case as an expert witness or not.

When you are called to be an expert witness, you'll have to have your CV in hand or already on file with the law firm you'll represent. You are there as an **individual on your own knowledge and skill of your profession** lending your expertise in a case that could decide the outcome. It may come out in a deposition where you work, but let them know up front that you are not representing the station in any way. If you wish to make it more formal, seek the advice of an attorney on what would be the best way to set up your business.

Every forensic meteorologist should have a working knowledge in the legal aspects of the court room and the process of trials from start to finish. For lawyers who are interested in knowing some of the internal workings of meteorology please reference (For Lawyers Only) at the end of this chapter.

Policy Statement on the Weather Service—Private Sector Roles (1991)

"The issue [of expert testimony for weather-related private litigation] is addressed in detail in Federal regulations (15 CFR parts

15a and 909.4) which state that NOAA employees will not provide such testimony and generally anticipate that the private sector will. However, exceptions exist where NOAA and the NWS could provide expert testimony, for example, in government-related cases. This, of course, in no way precludes the private weather industry's recognized role to provide expert testimony in both civil and government litigation."

Processes and Concepts in Civil and Criminal Litigation

*(*Randall B. Christison Attorney at Law as presented at the American Meteorological Society Conference of Forensic Meteorology)*

The Expansion of the Daubert Expert Evidence Doctrine Continues. Recent Decisions in Federal and Louisiana Courts

Few areas of law depend to a greater degree on expert testimony and evidence than litigation involving environmental law and toxic torts. Verdicts in such cases often turn on the credibility that the judge or jury affords a particular expert witness and his opinion. Thus, it is usually critical that a litigant in such a case have his own expert's opinion accepted by the court under the appropriate evidentiary standards, while at the same time arguing that the opposition's conflicting expert opinion does not meet the threshold of the standard for admissibility.

The importance of this evidentiary issue is amply reflected in the fact that, in the last six years, the United States Supreme Court has granted certiorari for appellate review of three cases that each deal with the standard of admissibility for expert testimony. To appreciate the significance of the high court's actions, it must be understood that the Court accepts relatively few of the many cases for which review is sought, choosing only those it considers of suitable import based on constitutional and policy concerns. Thus, the Court's trio of visitations in such a short time span to this evidentiary issue of the admissibility of expert opinion is remarkable.

Until 1993, the prevailing expert evidentiary standard was taken from the longstanding decision in *Frye v. United States*, 293 F. 1013 (D.C. Cir. 1923), which required that a scientific theory only be generally accepted in the relevant scientific community to serve as a basis for expert opinion. But with the adoption of the modern Federal Rules of Evidence in 1975, the Supreme Court stated that the focus of expert opinion was statutorily altered, warranting a different standard to test admissibility. In the case of *Daubert v. Merrell Dow Pharmaceuticals, Inc.*, 509 U.S. 579 (1993), the Court stated that federal courts should decide the issue of the admissibility of expert scientific opinion by the application of five factors:

(1) Whether the expert's theory has been or can be tested for validity;
(2) Whether the theory has been subject to peer review and publication;
(3) The known or potential rate of error when the theory is applied;
(4) The existence and maintenance of standards of control; and
(5) The degree to which the theory is accepted in the scientific community.

The federal trial courts are to apply these factors as a matter of a "gatekeeping" role to assure that expert opinion be both relevant and reliable to assist the trial court or jury in rendering a decision. While the court noted that the new rules of evidence are premised on a more liberal evidentiary thrust, these factors have served in practice under *Daubert* and the cases which have followed it to limit expert opinion to valid scientific theories and to weed out junk science.

After the *Daubert* decision, the Court considered as a separate issue in the case of *General Electric Company v. Joiner*, 522 U.S. 136 (1997), whether a different standard should apply in allowing expert opinion into the record as opposed to excluding admissibility. In that case, the lower circuit appeal court ruled that, while a trial court's decision to admit expert opinion is to be reviewed on the "abuse of discretion" standard, a court's **exclusion** of such evidence should be weighed against a "particularly stringent standard of review." However, the Supreme Court again emphasized that the trial court as "gatekeeper" is required to make the fundamental determination as to whether there is an adequate nexus between the data presented and the opinion posed, or instead is a connection borne only of the *"ipse dixit"* (he himself said it) of the expert's baseless assumptions. Whether the trial court admits or excludes the evidence in this role, the Court held that the decision is subject solely to the "abuse of discretion" standard as a matter of review.

The last case in the *"Daubert* trilogy" is the Court's recent decision in *Kumho Tire Co. v. Carmichael*, 1999 U.S. LEXIS 2189 (March 23, 1999). In that case, the issue was whether the *Daubert* factors are to be applied only in cases where expert opinion deals with purely scientific principles or, instead, are equally applicable in other areas of technical expertise. The expert's testimony in *Kumho* concerned a general theory as to the reason an automobile tire had failed, resulting in a severe accident. Using the *Daubert* analysis, the trial court excluded this form of engineering testimony as unreliable, but was reversed by the lower appellate court. The Supreme Court reinstated the trial court's determination, stating that the *Daubert* factors were equally applicable to engineering testimony - and to all forms of expert testimony—to the degree that each factor was relevant in considering the particular expert opinion.

Since the *Daubert* decision, the United States Fifth Circuit Court of Appeal, which reviews cases arising in Louisiana's federal district courts, has considered a number of cases in which it has applied the *Daubert* factors. Two of the most recent decisions are *Moore v. Ashland Chemical, Inc.*, 151 F.3d 269 (5th Cir. 1998) and *Black v. Food Lion, Inc.*, 171 F.3d 308 (5th Cir. 1999).

In the *Moore* case, a truck driver claimed to have suffered a permanent lung dysfunction due to short-term exposure to a toluene-based chemical spill. One of his expert physicians was to testify that the specific lung illness was a condition termed "reactive airways dysfunction syndrome." Though the physician expert was recognized as a respected and distinguished practitioner in the field of pulmonary medicine, he could not show a credible scientific link to toluene exposure and the diagnosed disease. While the Fifth Circuit Court panel of three judges reversed the trial court's exclusion of the evidence, the appeal court granted rehearing to consider the matter *en banc*. Relying heavily on the *Joiner* case, the court then held that the trial court had not abused its discretion in finding that the expert's opinion was unreliable, as based either on conjecture or inadequate evidence, in posing that the plaintiff's specific illness was linked to a short-term exposure to toluene.

Similarly, in the *Black* case, the plaintiff claimed that a slip and fall incident in defendant's store caused trauma resulting in the disease "fibromyalgia," characterized by generalized pain, poor sleep, loss of concentration, and chronic fatigue. As the *Kuhmo Tire* case had been decided by the Supreme Court upon hearing of this matter, the appeal court first noted that under that case the *Daubert* factors were unquestionably relevant to the admissibility of the expert physician's medical diagnosis. The magistrate judge at trial of a defense motion to exclude the diagnosis allowed the expert physician to offer her conclusion into evidence, but did not specifically apply the *Daubert* factors in more than a generalized manner. In reversing the magistrate's decision, the Fifth Circuit noted that the magistrate had not applied the factors with specificity to test the expert's physician's methodology of determining that the plaintiff's trauma could have caused the diagnosed condition. Though the physician attempted to rule out other causes for the disease, relevant medical literature indicated no causative link between trauma and contraction of plaintiff's condition. In the court's opinion, had the magistrate conducted a proper analysis under *Daubert* and *Khumo Tire*, "the utter lack of any medical reliability of [the physician expert's] opinion would have been quickly exposed."

Lastly, and most significant in litigation in the state's courts, the Louisiana Supreme Court, in *State v. Foret*, 628 So. 2d 1116 (La. 1993), adopted the *Daubert* standard for testing expert testimony nearly immediately after it was articulated. Noting that Sections 702 of both the federal and the Louisiana evidence codes pertaining to the admissibility of expert opinion are identical, the state supreme court stated that it should follow *Daubert* on the same reasons stated therein. In *Foret*, the criminal defendant had been convicted of sexual abuse of a minor, based in part on the expert testimony of a child psychologist. The psychologist used an interview technique, developed for diagnosing the effects of child abuse to prescribe appropriate treatment, as a basis for an opinion that the victim's interview responses positively indicated she had been abused. But, the court cited to literature that stated the technique was unreliable in delineating actual abuse from fantasy or fraud, and was accepted as valid only for its intended purpose. Based on the *Daubert* factors, while the court did not exclude the testimony of its own accord, it remanded the case for a new trial with the admonition that the trial court exercise its "gatekeeping" authority to closely evaluate any such testimony offered at retrial.

Daubert has been cited in reported Louisiana state court cases dozens of times. The most recent citation is in *State v. George*, 1999 LEXIS 447 (La. Feb. 26, 1999), where the court sided with the trial court's exercise of the "gatekeeping" function to exclude evidence of defendant's alleged "limbic psychotic trigger reaction," asserted as a justification for violent conduct.

As seen in the foregoing cases, the admissibility or exclusion of expert opinion after *Daubert* can be decisive in an environmental or toxic tort case. Of significance to Louisiana businesses, a bill currently is being considered in the legislature (HB 247) which would impose a "loser pays" rule as to legal fees and costs against plaintiffs who bring frivolous lawsuits. The above decisions indicate that a challenge to the expert basis of a case under the *Daubert* factors can be key in exposing baseless litigation. Should the legislature pass this bill, the "*Daubert* challenge" may well be one method used to test the relative merits of a matter alleged frivolous by the defendant.

Courtesy: M. G. Durand of *Onebane Law Firm*

Often there are times when black ice becomes an issue in a defense. "Sudden emergency" as lawyers would describe it.

Defendant's admitted knowledge of worsening weather conditions and presence of road hazards does not justify charging the jury with an "emergency doctrine" instruction. What follows is a case from the New York State Court of Appeals decision in the *Caristo v. Sanzone* matter.

Cases involving "Sudden Emergency" Defense

New York Court Opinions
Antoinette Caristo et al.,
Appellants,
v.
Augustine Sanzone et al.,
Respondents.

2001 NY Int. 37
April 3, 2001
This opinion is uncorrected and subject to revision before publication in the New York Reports.
Arnold E. DiJoseph III, for appellants.
Alan M. McLaughlin, for respondents. GRAFFEO, J.: The issue in this motor vehicle accident negligence case is whether the trial court erred in charging the jury on the emergency doctrine. Under the facts presented, we conclude that defendants were not entitled to this instruction.

At approximately 9:00 A.M. on the morning of the accident, defendant Augustine Sanzone was driving a vehicle owned by his wife, defendant Patricia Cinquemani, on Foster Road in Staten Island. At the same time, plaintiff Antoinette Caristo was operating her automobile on Woodrow Road. Foster Road terminated at a "T" intersection with Woodrow Road, and a stop sign controlled the flow of traffic from Foster Road onto Woodrow Road.

At trial, Sanzone testified that the weather conditions at 7:00 A.M. that day consisted of snow, rain and freezing rain. This mixed precipitation was unchanged when he and his family left their home at approximately 8:30 A.M. By the time he drove to Foster Road, the weather had worsened. He described the conditions as "more like frozen rain and hail at the time." The temperature that morning was established, by stipulation of the parties, at 22 degrees Fahrenheit.

After cresting an incline on Foster Road, Sanzone proceeded downhill toward the Woodrow Road intersection, traveling at 20 to 25 miles per hour. At this juncture, his vehicle began to slide and he noticed there was "a sheet of ice" on the hill. Despite Sanzone's effort to pump the brakes, the vehicle slid 175 to 200 feet, past the stop sign and into Woodrow Road. As plaintiff approached the intersection at 15 to 20 miles per hour and saw defendants' vehicle, she attempted to swerve to avoid a collision, but was unsuccessful. Both Cinquemani and the police officer who responded to the scene of the accident confirmed the icy conditions on Foster Road. Neither plaintiff nor Sanzone experienced difficulty controlling their vehicles prior to this incident. Over plaintiff's objection, the trial court charged the jury on the emergency doctrine. The jury returned a verdict in favor of defendants and the trial court entered a judgment dismissing plaintiff's complaint. The Appellate Division affirmed the judgment, with two Justices dissenting (274 2 406). Plaintiff now appeals as a matter of right.

More than a century ago, this Court first considered the reasonableness of an actor's conduct when confronted with a sudden emergency situation (*see*, *Wynn v C.P., N.&E. R.R.R. Co.*, 133 NY 575). Since then, we have articulated and applied the common-law emergency doctrine which "recognizes that when an actor is faced with a sudden and unexpected circumstance which leaves little or no time for thought, deliberation or consideration, or causes the actor to be reasonably so disturbed that the actor must make a speedy decision without weighing alternative courses of conduct, the actor may not be negligent if the actions taken are reasonable and prudent in the emergency context" (Rivera v New York City Tr. Auth.,, 77 NY2d 322, 327), provided the actor has not created the emergency.

The rationale for this doctrine—the need to instruct a jury that it may consider the reasonableness of a party's conduct in light of the unexpected emergency confronting that person—has been somewhat eroded by the evolution from contributory negligence to comparative negligence. With the advent of the ability of juries to allocate fault and apportion damages, the viability of the doctrine has been questioned by some jurisdictions, with a few states going so far as to abolish it (*see generally*, Modern Status Of Sudden Emergency Doctrine, Ann. 10 ALR5th 680).

In New York, in addition to the elements of the charge, we have defined the role of the Trial Judge in assessing the propriety of an emergency charge request. We require the Judge to make the threshold determination that there is some reasonable view of the evidence supporting the occurrence of a "qualifying emergency" (Rivera v New York City Tr. Auth., supra, 77 NY2d, at 327). Only then is a jury instructed to consider whether a defendant was faced with a sudden and unforeseen emergency not of the actor's own making and, if so, whether defendant's response to the situation was that of a reasonably prudent person (*see*, PJI 2:14).

The emergency instruction is, therefore, properly charged where the evidence supports a finding that the party requesting the charge was confronted by "a sudden and unexpected circumstance which leaves little or no time for thought, deliberation or consideration" (*Rivera v New York City Tr. Auth.*, supra, 77 NY2d, at 327; Kuci v Manhattan & Bronx Surface Tr. Operating Auth.,, 88 NY2d 923, 924; *see also*, Restatement [Second] Torts § 296). Here, even considering the evidence in a light most favorable to defendant (*see*, Kuci v Manhattan & Bronx Surface Tr. Operating Auth., supra, 88 NY2d, at 924), we hold as a matter of law that there was no qualifying event which justified issuance of the emergency instruction. Given Sanzone's admitted knowledge of the worsening weather conditions, the presence of ice on the hill cannot be deemed a sudden and unexpected emergency.

Although Sanzone did not encounter patches of ice on the roadways before losing control of his vehicle, at the time of the accident the temperature was well below freezing and it had been snowing, raining and hailing for at least two hours. As such, there was no reasonable view of the evidence that would lead to the conclusion that the ice and slippery road conditions on the Foster Road slope were sudden and unforeseen. Defendants were not, therefore, entitled to an emergency instruction and the charge to the jury constituted reversible error under these circumstances.

The dissent contrasts our holding here with *Ferrer v Harris* (55 2 285), where we concluded defendant was entitled to an emergency doctrine charge. *Ferrer* is clearly distinguishable in that defendant was confronted by an unanticipated event when a four-year old child ran in front of his vehicle from between two parked cars. The qualifying emergency—a child darting from a sidewalk into street traffic—is simply not analogous to the presence of ice and slippery conditions following at least two hours of inclement weather with temperatures well below freezing. Accordingly, the order of the Appellate Division should be reversed, with costs, and a new trial ordered. Caristo v Sanzone No. 22 Rosenblatt, J. (dissenting): By holding it was error to charge the emergency doctrine, the majority concludes *as a matter of law* that defendant expected to encounter a 175 to 200 foot sheet of ice while driving on an otherwise ice free road. In my view, it was for the jury, and not the court, to determine whether this sheet of ice was "unforeseen." I agree that the emergency doctrine should not be charged merely for the asking; nor should it be charged in every foul weather fender-bender case. But here the charge was justified. Accordingly, I dissent.

Because we are required to view the evidence most favorably toward the party requesting the emergency instruction (*see, Kuci v Manhattan & Bronx Surface Tr. Operating Auth.*, 88 NY2d 923, 924), we must accept the following: First, defendant drove onto a sheet of ice 175 to 200 feet long. Second, the character of the ice was such that, despite his efforts to stop the car, he slid the entire 175 to 200 foot distance—a journey approximately two-thirds the length of a football field. Third, defendant was unable to see the sheet of ice until he was upon it. Fourth, defendant was driving only 20 miles per hour when he encountered the ice. Fifth, although the weather was bad (rain, snow, sleet and freezing temperatures), neither plaintiff nor defendant had encountered ice in the vicinity. In my view, this raises a question of fact as to whether defendant was confronted with an emergency.

It is settled law that an emergency doctrine charge must be given if, under some reasonable view of the evidence, the party requesting it was confronted with a "qualifying emergency" at the time of the alleged tortious conduct (Rivera v New York City Transit Auth., 77 NY2d 322, 327). A "qualifying emergency" is a "sudden and unforeseen occurrence not of the actor's own making" (Rivera, supra, 77 NY2d, at 327 [citing *Ferrer v Harris*, 55 NY2d 285]). Given these facts, the jury could reasonably have concluded—as they obviously did—that encountering this enormous sheet of ice was "a sudden and unforeseen occurrence." Over the course of a lifetime, few winter drivers will ever encounter an ice hazard of that magnitude—let alone when the roads are otherwise free of ice. Accordingly, the trial court properly gave the charge. In doing so, it did not conclude that an emergency existed. It merely ruled that, on the evidence presented, the existence of an emergency was debatable and allowed the jury to resolve the point.[1] Charging the emergency doctrine simply reminds the jury that it must consider the reasonableness of a party's actions in light of the existing circumstances (*see*, *Ferrer v Harris*, supra,, 55 NY2d 285, 292 (citing *Wagner v International Ry. Co.*, 232 NY 176, 182 [Cardozo, J.]).[2] As we noted in *Rivera*, giving the emergency doctrine instruction is "by no means" a directed verdict for the party requesting it (*Rivera*, supra, at 435) or even a declaration that there was an emergency. Rather, the jury still has the final say as to whether there actually was an emergency and, if so, whether the party reacted to it reasonably (*cf.*, *Rodriquez v New York State Thruway Auth.*, 82 AD2d 853, 854 [party's actions still unreasonable even though he was confronted with an emergency]).

In *Ferrer v Harris* (supra,, 55 NY2d 285, 290–291), defendant was driving his car 20 miles per hour on a block filled with children. Plaintiff, a young girl, darted into the street and was struck by defendant's car. Defendant testified that he saw children playing on the sidewalk before the accident. He even admitted seeing plaintiff step between two parked cars seconds before she darted (*Ferrer v Harris*, supra, 55 NY2d, at 290–291). The trial court refused the emergency doctrine charge. We held that the trial court's *failure* to charge it was reversible error because "it was more than conceivable that a jury could conclude that *this* defendant was faced with an emergency" (*see*, *Ferrer v Harris*, supra, at 292 [emphasis added]).

More recently, in *Rivera*, we again reversed for *failure* to give the emergency doctrine charge. We concluded that "the jury could reasonably have concluded" that the accident was sudden and unexpected (*see*, *Rivera*, supra, 77 NY2d, at 327). Indeed, we have consistently authorized if not required the charge so as to allow the jury to resolve whether an emergency occurred and whether the party's actions were reasonable in the face of it (*see*, *Kuci v Manhattan & Bronx Surface Tr. Operating Auth.*, supra,, 88 NY2d 923, 924 [trial court committed reversible error by failing to charge emergency doctrine because driver testified that he "did not anticipate being suddenly cut off by this particular car"]; *Mas v Two Bridges Assoc.*, 75 NY2d 680, 686 [upholding trial court's emergency doctrine charge because "we think that on the evidence in this case the emergency was not dissipated * * * as a matter of law, and that the question was properly submitted to the jury"]; *Amaro v City of New York*, 40 NY2d 30, 37 [emergency charge was properly given]).

> Referenced from the New York Court of Appeals, *CARISTO v. SANZONE*, April 3, 2001. Presiding Judge: Judge Graffeo

I recognize that retention of the emergency doctrine has been under discussion nationally and that some jurisdictions have altered or abolished it.[3] The doctrine, however, is still a part of New York law. That being so, I submit that the trial court and the Appellate Division correctly applied it. The trial court and the Appellate Division majority concluded that the matter was at least arguable, which is to say that there was "some reasonable view of the evidence" that defendant was confronted with an emergency. The jury went even beyond that; as fact-finders they concluded that defendant was indeed confronted with an emergency and his actions were reasonable in light of it. Their verdict does not strike me as irrational.

Accordingly, I would affirm the order of the Appellate Division.

Order reversed, with costs, and a new trial ordered. Opinion by Judge Graffeo. Chief Judge Kaye and Judges Levine, Ciparick and Wesley concur. Judge Rosenblatt dissents and votes to affirm in an opinion in which Judge Smith concurs.

Litigation Proceedings

Listed below are types of Litigation Proceedings and categories that were presented by Sean Potter, CCM at the AMS Conference Workshop for Forensic Meteorologist, San Diego, CA 2005. Additions were added by the author. Because of the vast examples these categories could play, we have elected not to describe them here.

Civil

- Personal Injury
- Hourly Weather Data (Temperatures)
- Snow and Ice accumulation
- Visibility
- High winds
- Property Damage

    - Wind data
    - Rainfall records
    - Thunderstorms
    - Lightning
    - Hail

- Motor Vehicle Accidents

    - Rainfall, snowfall
    - Visibility (Restrictions to visibility)
    - High winds (Wind Shear)
    - Illumination data (Sun/Moon Data)
    - Cloud cover

- General and Commercial Aviation

    - Aircraft icing
    - Turbulence
    - Wind Shear
    - Thunderstorms
    - Micro bursts
    - Temperature (Warmer air requires longer take-off distance) (Overloaded)

- **Maritime**

  - Storms
  - Winds
  - Wave heights
  - Visibility

- **Precipitation (Ice in winter, Hail in summer)**

Construction Delays (Planning will be defined under Engineering—in a different chapter)

- **Thunderstorms/Lightning**
- **Strong winds**
- **Heavy rain/snow**
- **Flooding**
- **Temperature (Restricting certain construction procedures)**

Air Pollution

- **Dispersion Models**
- **Wind profile (Vertical and horizontal)**
- **Stability**
- **Toxic Corridor (Chemical dispersion/rate/amount/severity)**
- **Criminal**

Investigations

- Insurance

  o False claims filed
  o Verification of damage associated with storms watches/warnings, black ice (sudden defense).
  o Some cases may be investigated by insurance companies without ever seeing the inside of a court-room.

- Environmental Impact Assessment

Legal Jargon and Definitions:

- Parties: In civil cases, the Plaintiff is the person or entity that filed the lawsuit and the Defendant is the person or entity being sued.
- Expert witness vs. Percipient Witness: Experts can rely upon hearsay information, and may express opinions based upon their training and expertise. Percipient witnesses can only testify to what they have seen, heard, etc.
- Discovery: Expert Deposition. A process where the expert witness gives testimony, under oath, regarding their expert opinions and the reasons for the opinions.
- Trial Participants:
- Judge: The person who presides over the trial, and makes legal rulings on issues that arise during the case.
- Marshal: The law enforcement person who maintains security.
- Clerk: Is the person who handles exhibits, pleadings and calendaring in the court room.
- Reporter: Is the person who prepares a shorthand record of all trial testimony.
- Jury: The individuals approved by both Plaintiff and Defendant Attorneys who decide the case.
- Voir Dire: The process by which judges and lawyers select a petit jury from among those eligible to serve, by questioning them to determine knowledge of the facts of the case and a willingness to decide the case only on the evidence presented in court. "Voir dire" is a phrase meaning "to speak the truth."
- Hypothetical Question: A question that sets forth certain assumptions, and asks for an expert opinion based upon those assumptions.
- Negligence: Doing something that a reasonable person would not do, or not doing something that a reasonable person would do.
- Reasonable scientific probability: What is required for a scientific expert opinion?

- Beyond a Reasonable Doubt: The burden of proof for the prosecution in a criminal case.
- Preponderance of the Evidence: The burden of proof for the plaintiff in a civil case.
- "Junk" science: Theories that are not supported by relevant scientific literature and practitioners.
- Verdict: The decision made by the jury. This is later turned into a judgment.

Opinion Testimony & Scientific Evidence

(*Karen Holmes Balestreri, Pendleton & Potocki as presented at the AMS Conference of Forensic Meteorologist. Jan/2005)

A. California Evidence Code §801

If a witness is testifying as an expert, his testimony in the form of an opinion is limited to such an opinion as is:

(a) Related to a subject that is sufficiently beyond common experience that the opinion of an expert would assist the trier of fact; and
(b) Based on matter (including his special knowledge, skill, experience, training, and education) perceived by or personally known to the witness or made known to him at or before the hearing, whether or not admissible, that is of a type that reasonably may be relied upon by an expert in forming an opinion upon the subject to which his testimony relates, unless an expert is precluded by law from using such matter as a basis for his opinion. Leg.H. 1965 ch. 299, operative January 1, 1967.

B. Federal Rules of Evidence §702

If scientific, technical, or other specialized knowledge will assist the trier of fact to understand the evidence or to determine a fact in issue, a witness qualified as an expert by knowledge, skill, experience, training, or education, may testify thereto in the form of an opinion or otherwise, if (1) the testimony is based upon sufficient facts or data, (2) the testimony is the product of reliable principles and methods, and (3) the witness has applied the principles and methods reliably to the facts of the case.

I should mention this simply because it has helped me tremendously and that's having the advice of a lawyer who is looking out for your best interest. I'm not talking about the lawyer your representing as an expert witness; I'm talking about someone who is on the outside looking in. There are a number of legal benefit companies that offer such services but you usually need to be employed by a company in order to receive such benefits. One such company Pre-Paid Legal Services offers such a service to the individual. I only pay $25/month and I can call my lawyer on any issue. If I'm worried about a conflict of interest, what my liabilities are, risks involved . . . I have my bases covered. Plus, they're services go far beyond consulting which are included in that monthly fee. Contact *http://www.WLSorg.com/* to find out more. Or you can contact Pre-Paid Legal directly at *www.pplsi.com* .

There are other companies out there and I'm not familiar with them as I have only used this particular service. Feel free to shop around.

Sample Expert Testimony Letter

Name
Title
Business
Street
City, State, Zip
Phone
E-mail

Date:

Lawyer
Address
City, State Zip

RE: Plaintiff v. Defendant

Dear Mr. Lawyer,

Per your instructions, I have reviewed the necessary meteorological records, supplemental information from local, state and governmental authorities & coop weather observers that surrounds the area (accident) and am submitting my conclusions for your review.

Overview:

You asked about the weather and surface conditions that led up to the date of the accident—in this case—a slip and fall in the parking lot, on December 24th, 2001 at a business called XX. You asked me to look into the possibility of ice forming prior to the snowfall during the early morning hours of December 24th, which could have led to the cause of the accident.

Data:

The amount of official weather data from observation locations is limited due to the population of (region). Considering the close proximity of the business to the local airport, data from the ASOS (Automated Surface Observation System) was used. It would be inappropriate to use data from other regional ASOS stations as the micro-climate of (Region) can vary tremendously, especially when locations are close to a large body of water (in this case, XX Lake). I reviewed the observations from the airport as well as local Department of Transportation logs showing county snowfall as well as the snowplow log from the company hired to maintain said parking lot.

Discussion:

This where you would get "down and dirty" with your reasoning and theory. Using all data available and putting it in a logical, easy to understand format. No junk science!

I added one of the formulas I have used in past cases:

Using Stefan-Boltzmann Law: If object at temperature T, in this case, Asphalt, is surrounded by an environment (air) at temperature T_0, the net heat flow is:

$$H_{net} = e A [T^4 - T_0^4]$$

This is why the onset of snow would melt when making contact with asphalt.

The opposing expert testimony would only be correct for a lot with no drainage (standing water) and conditions where the atmosphere is in (a state of equilibrium) where no mixing (evaporation/sublimation) would occur. This (state of equilibrium) is theoretically, scientifically and operationally impossible as there's always some form of energy flux within the atmosphere, especially given the strength of the surface wind flow and drastic change in temperature (as indicated by surface observations) to induce sublimation.

Supporting Documentation:

Plow records from County (MDOT)
Snowplow billing record from XX
Weather records for XX Airport for December 23 and December 24th, 200X
Pictures showing conditions before rain, during and after rain. Date and times and current weather noted at bottom of pictures. (Note: Sufficient draining)

Conclusions:

Awaiting info from XXDOT about drainage info . . . . .

I certify that the above information is true and accurate and that any estimations, interpolations or assumptions that have been made were done so with expert accuracy by a professional meteorologist. Additionally, I reserve the right to amend these conclusions made herein upon further discovery of meteorological data.

Sincerely,
Name
Consulting Meteorologist
XX Company

The suggested format is merely a sample as you will want to change or modify to suit your needs. But it's a good start if you have nothing to reference from. As your knowledge grows and you develop more cases, you'll develop a laundry list of additional comments, reference to manuals, operating procedures and map/graphics/pictures. Notice the last paragraph:

> *"I certify that the above information is true and accurate and that any estimations, interpolations or assumptions that have been made were done so with expert accuracy by a professional meteorologist. Additionally, I reserve the right to amend these conclusions made herein upon further discovery of meteorological data."*

If you come across information that could sway your testimony, you have the right to make changes. Don't hold back information that could sway the case, you need to be honest, ethically and morally correct. Words like "possible" or "alleged" are not words that should be used in testimony. Be accurate and have sound theoretical, mathematical, scientific or operational reasoning backing your testimony. It's your reputation that's on the line.

What you need to get started.

(Parts presented by Jan Null, CCM, AMS
Conference Forensic Workshop,
San Diego, CA 2005)

Before the phone rings, you need to be ready to provide the necessary documents that will define your ability to work as a forensic meteorologist. Here's a list (basic) that you need to keep current:

- Resume
- CV (Circum Vitae)
- Fee Schedule

    o Rate Structure
    o Travel Expenses
    o Billing

A resume is needed to define your accomplishments based on years of experience. Don't overstate your qualifications! That's the quickest way for the opposing counsel to shoot holes in your testimony.

Jan Null, CCM says "A CV is not necessarily a shorter version. It just has a different focus and leaves out a "career goals" section. A good overview can be found at: *http://www.brandeis.edu/hiatt/cv.html*"

Usually list your specific reports/papers that were peer reviewed; manuscript(s) that was published in professional magazines and such. List speaking engagements and previous cases that you worked that defined a new law. Information that shows credibility in your current profession should be listed.

Fee Schedule varies depending on location. Check your other competitors or ask a lawyer in your area what feels adequate for the area. Here is a **sample** Fee Schedule:

Phone conversation:	$10 per 6 minutes	$60 /hr
Drafts of information	(Simple review)	$75
Data request	NCDC	$90
Research		$120/ hr
Analysis (Micro, Macro)		$150/hr
Meeting with Counsel		$90/hr
Final Expert Witness Document		$100
Depositions		$350/hr (2 hr minimum)
Called to Testify		$500/hr (2 hr minimum)
Time to & from scene		$90/hr
Travel expenses (food, lodging, airfare)		
Billable from receipts.		

This is best calculated using a spreadsheet (Microsoft Excel) or a similar program that has it available to fax, e-mail/present to your requesting party. You may find some charging a flat rate fee by the hour (portal to portal) from $375/hr to as high as $1100/hr.

As you compare this schedule to other professionals in the same field or your own, you will see that these are below the standards.

Billing should accurate with a complete list of your time and how it was divided. Don't pad or over bill. Keep an accurate record of your work.

It wouldn't hurt if you made an effort to meet some of the lawyers/law firms in your area. Have business cards, a copy of your CV, Fee Schedule and ask what cases they may be working on at that time that may be questionable as to whether the environment played a role in the accident. Advertise in their quarterly bulletin. Numerous lawyers are unaware as to how the weather can affect a crime scene or accident. I'll touch on this later in this chapter.

So you've received your first call about weather information. Did you get all the pertinent information about the case? Having a preprinted contact sheet with the basics would help. Here's a list of the basics;

Firm Name
Name of Attorney
Firm Address
Phone number
Fax
E-mail
Case# (or File #)
Plaintiff v. Defendant involved:
Exact Time and Date of Incident
Location of Incident
Ask what pertinent information they are requesting.
Deadline

If you are called as an expert witness, you may simply get a call stating that you have been added on their case. Be sure to log that information as you may receive another call from the opposing law wishing to use you as their expert witness. That would be a conflict of interest. Log each call by "Case/file" number or "Plaintiff/Defendant" or "Jurisdiction Court".

With all the necessary information in hand, you now are on your way to your first of many cases in your career. Your list of resources will determine how long it takes for you to gather vital information to your case. Get to know anybody who has an interest in weather. Farmers who keep daily records, plow companies, harbors, anybody! This will come in handy when the time comes where you need to cross reference your weather data from a distant recording station. Try to use public data that you know has been calibrated or quality controlled. Some sources would be the US Forest Service, National Weather Service, local water districts and the data should be freely available (not to say it's free—but available to both parties).

I had attended a business expo in my area and met with a number of businesses as well as some law firms who were represented by local counsel. This is a great way to drum up business! One of the lawyers asked what I did and I told him I was a forensic meteorologist. He looked puzzled and asked what that was. After a quick summary and sample cases I worked, he was surprised and then said that his firm doesn't need the advice of a forensic meteorologist. After we exchanged business cards, I realized his downstate counterpart was the opposing counsel on a case I was called to testify on. It was an automobile accident involving snow, ice and wind shear. I asked him how that case was decided (knowing full well how it ended), he said that he wasn't sure but that it may have been dismissed (they lost their case). I told him I was the expert witness hired by the defendants' counsel. His facial expression was priceless! I shook his hand and bid him a good day.

As a forensic meteorologist, you need to have the foresight to see how the weather can play a factor and explain that to the counsel. Know your formulas, thermodynamic principles and be ready to learn parts of other professions.

The Process

Initial contact

Usually by letter, phone call or e-mail asking for information. This is also the time where you should log the contact and deter anyone else from hiring your services since they were the first to contact you, as this would be a conflict of interest.

Phone interview

Usually to review the "meat" of the case and what they're looking to do.

Face-to-face meeting to go over litigation and strategy:

Usually conducted in the lawyer's office or conference room depending on what part of the nation you're in. Review all documents pertaining to the case, previous depositions, testimony, review citations if available and formulate a plan of strategy.

Citations from various law enforcement departments can vary greatly. It's this variance that can often create holes in a testimony and allow the opposing or defense team to pursue legal action. The more weather information that can be recorded at the crash/crime scene, the margin of error is reduced.

Expert's research:

Forensics is: **Fact based precision documentation.** Examine the opposing expert witness testimony (if available). Conduct your own research, visit the site, gather data and start formulating a rough draft of how you want to prepare your thoughts for the report.

Preparation of expert report:

Due to discovery rules, find out if your attorney wants any written work, including correspondence by e-mail. A sample report is in this chapter which gives a basic outline how to proceed and the format that is generally accepted in a court of law.

Review of report with counsel, possible revision:

Once you have all the materials, copies of data, pictures available. Present this to your counsel for a review to see if it meets the criteria they are using in a defense case (or Plaintiff's case). Some reports may be no more than a few pages and I have had a few that were more than 150 pages. It all depends on the supporting documentation and evidence available.

Preparation for deposition:

After your report has been viewed by the opposing counsel, they may want to take your deposition. This is a process where the opposing lawyer (or more) will want to verify your credentials and try to bring discredit to your testimony. This is not the time to fall apart! Keep your cool and make sure you have analyzed your investigation from BOTH sides. If you're hired by the defense team, take time to review your report as if you were on the plaintiff's team. Look for mistakes; any documentation that you can't support through solid mathematical formulas, empirical formulas, operational analysis or weather fundamentals, shouldn't be in the report. Either find a solution or don't use it. No junk science!

Deposition process

As stated above, this could take an hour, or some can go as long as 6 hours. It all depends on how complicated the case is and how many are named in the lawsuit(s). Keep your cool! There will be times when you go up against lawyers who have no knowledge in weather. Questions almost seem to be stupid in nature, but they don't understand the atmosphere like we do. I have asked the stenographic to stop "can we go off the record for a moment please?" so I can give a crash course in Meteorology 101. Showing how the weather works, (for instance, why is a prevailing westerly wind from the west) levels the playing field for everyone at the table and then you can continue on. Some depositions have been very formal with "yes" and "no" answers, others have been like a classroom environment where I was teaching more than I was answering. You'll get all kinds of questions, some very direct and some extremely vague. Also expect them to question you five different ways on the same topic. It's the job of the attorney who is deposing you to find holes in your testimony, so be prepared to be tested.

Be professional, courteous and well-dressed; have all of your notes available. Leave nothing to chance.

Tips on testimony:

> Direct answers only; don't volunteer unsolicited information from opposing counsel, consistency of professional opinions, etc.

Review and editing of stenographic deposition record "Woodshedding" (pre-trial preparation)

The chances of it going to trial are slim. But, just in case, you'll have plenty of time to review your report with counsel and review the depositions of all parties involved. By the time the case does go to trial, be sure to review them again as this process can take a long time and you could have very well provided a dozen or so cases.

Courtroom procedure and decorum

The trial may take anywhere from a day to week. Your lawyer will ask you to be available on certain days but won't be able to pin down an exact time because the trial is constantly in flux. So have your schedule open on those days. You'll most likely be asked to sit in another room or outside the courtroom proceedings until you are called to testify. When called, follow your lawyer's advice and directions.

Just like the deposition, come prepared with the materials you used to come to your conclusion. If you forget to bring a few things, no problem, you can ask them to jog your memory on some exhibits that have been admitted as evidence. Anything else you bring can be admitted as discovery evidence (even a woman's purse). Your presentation style may vary, it could be as simply as a piece of paper to a Power-Point presentation; whatever you need to support your testimony.

Be polite and professional and should your cross-examination from the opposing counsel get under your skin, don't let it bother you. That's their job! You may pick up on a question asked five different times in five different ways. The answer will be the same. They're there to trip you up. Be patient and calm and be a professional.

This may shock you but the way you present yourself and what you wear has a deciding factor if the jury is going to like you. Body piercing, tattoos and the like are not going to fare well with the jury. You need to be professional, a suit for the man and a skirt below the knees for the women. For the man, do not wear a red tie as this shows a sign of aggression. Keep it a color that is consistent with the rest of your suit. For the women, wear your hair up. Again, look professional.

For Lawyers Only

Every case has its uniqueness, yet there are similarities in most winter accidents and that's ice, black ice (man made), packed snow or wind shear (at least in my cases anyway). There's one element of my investigation that has always assured a success in the end and that's scene evaluation. Either by pictures, (aerial or preferably ground), topography charts visual means. If your expert witness is local, this shouldn't be a problem. However, if you're seeking someone who works nationally and can't be at the scene, it would be a good idea to offer up some pictures, video or something that can give him/her a better idea of the scene of the accident. For without this, testimony may be incomplete or flawed.

Sudden Emergency, provided by Mother Nature or **man made**. That's right! man made. Here's how it works.

[**Black ice** is a thin formation of ice that occurs when water freezes on a road surface. Because it is so thin and clear, the dark underlying road surface easily shows through, lending to its common name. Black ice can naturally form at any temperature at or below freezing, but it is favored to form when the road surface is between 25F and 32F (-4C and 0C). This range was observed on February 10th in the late evening hours during snowfall.

A deicing compound, salt (Sodium Chloride), magnesium, etc. is applied in a concentration or mixture that will melt ice or snow on the highway at temperatures which may be much below the normal range of black ice formation. The applied compound starts melting the packed snow or ice on the highway. But, if the temperature does not rise, or worse, falls, the compound, diluted by the melting action is now below its working temperature and will start to freeze. The Department of Transportation will scrape/plow the roads and apply salt when accumulated snow is at 1 inch.

If left unattended and more concentrated deicing agent is not immediately applied to strengthen the dilution back to a point where it again melts, "black ice" forms on the roadway. In the event the temperature falls and no precipitation occurs, (and in this case, the temperature dropped to the low 20's) the lower temperature may now be below the liquid point of the applied deicing agent, which would result in a thin layer of frozen or slippery 'deicing' compound that is now 'black ice'. If the humidity drops to a level where equilibrium with the saturated solution is reached, and the humidity continues to drop further for a longer period of time, (2–5 hours) [Note the drop in relative humidity prior to the accident] a portion of the calcium chloride solution on the road is converted into a solid form, (hexahydrate - $CaCl_2 - 6 H_2O$). This causes "ice to develop."

This can be a very dangerous condition because where there was packed snow or traces of ice and snow that were visible to the driver, this has now dissolved into a thin liquid film that refreezes and is not readily seen, and is unexpected at the low temperatures! This is more prevalent in and around intersections. The continuous snowfall with periodic wind gusts to 15 mph suggest that there was minor blowing snow over the road where the roads was not protected.

(From the files of Greg MacMaster, Forensic Meteorologist)

More food for thought: accidents on asphalt where the air temperature cooled below 32 degrees and visible moisture was present would give a perfect case for ice to occur. In my research for the cooling aspects of asphalt, I came a cross a computer program called PaveCool. The programmer had to modify it for temperatures below freezing, but it did the job and our defense team was happy with the court ruling. PaveCool shows the delay of cooling based on many factors, one being air temperature, thickness of the asphalt, grade and wind speed. In one case I found that the surface temperature of the asphalt cooled slower than the air, which delayed the onset of ice—by 3.4 hours. To access this program—go to:

> *http://groups.yahoo.com/group/forensic_meteorology/* or contact the author at *Cyclogenesis@charter.net* .

More tidbits of information: drainage of parking lot or roadway, anything to show the drainage capability of the scene of the accident. Too many testimonial reports are fixed on a fact that isn't true in most cases. Christopher Columbus found out that the world wasn't flat and neither is a parking lot. State construction guidelines dictate the amount of slope (usually 1/100) or 1 foot of drop for 100 foot of run. Anything greater than that and you may not have standing water freezing!

In winter, salt/calcium chloride is often distributed on main roads to keep them from freezing. As vehicles drive in driveways, parking lots and other areas not usually treated, they'll usually carry a certain amount of salt/calcium chloride into the area. Although it's diluted, it can be enough to aid in a delayed onset of ice.

Now I'm not trying to give a defense to all your cases, just broadening your scope of thinking in an altered environment. Weather isn't always black and white, but a gray scale. We have mathematical formulas, hypothetical reasons and operationally proven conclusions and if the right testimony is provided, there should be no margin for error and should be iron clad. The items I detailed above are more for the defense counsel. We're in a litigious society and a good defense beats any offense.

I have reviewed testimony from a fellow professional and it would seem to reason that there would be a line of reasoning consistent with each other. Sometimes that's not the case. You may receive an expert witness testimony that describes exactly what you're looking for. You need to be careful as you'll open yourself to a resounding display of opposing documents that could refute your original testimony and place doubt in your mind as to his/her abilities. Fairness, honesty and integrity should be the motto of forensic meteorologists and I'm hoping this book will reinforce this.

Engineering Physics for Lawyers

I. INTRODUCTION:

Much of the work of accident analysis and reconstruction involves the application of general physical laws to the particular situation at hand. These physical laws are expressed in terms of Newtonian mechanics. Newtonian mechanics is the base science of physics and engineering. Today the development and application of basic Newtonian mechanics is almost exclusively the province of engineers, in that physicists, for the most part, conduct basic research only in "new physics" areas such as quantum mechanics, relativistic physics, etc.

II. KINEMATICS: THE ANALYSIS OF MOTION

    1. **Velocity.** Velocity is the rate of change of the position of a body over time (velocity = distance/time). Velocity is a vector, which means it has both a magnitude and direction. Thus, 30 miles per hour north, or 20 meters per second along the X axis, are both velocities. Note that the more common term, speed, is a velocity without a direction, so that 30 miles per hour or 20 meters per second are speeds, and not velocities, since no direction is specified.

    2. **Acceleration.** Acceleration is the rate of change in velocity (acceleration = velocity/time). Thus, any time a body changes its rate of travel or its direction of travel, it is said to be accelerating. Acceleration is a vector

requiring both magnitude and direction. Thus, for example, 32.2 ft/sec/sec. downward toward the center of the earth, is the acceleration vector of all bodies on the surface of the earth. If the support collapses, for example, if a person falls off a ladder, they will increase their velocity at a rate of 32.2 ft/sec. per second. At the end of one second of free fall, they will be traveling at a rate of 32.2 ft/sec. which is approximately 22 miles per hour. They will have fallen approximately 16.1 feet in this one-second interval. (D = 1/2 a t x t = 1/2 x 32.2 x 1 x 1 = 16.1).

III. NEWTON'S LAWS

Newton wrote three laws of motion which relate kinematic phenomena to something else—forces. These are:

1. *First Law:* "A body does not alter its state of motion without the influence of an external force." That is, there is no change in the velocity of a body (neither in magnitude nor in direction) unless some force acts on that body.

2. *Second Law:* "The net resultant force applied to the body is equal to the first time-derivative of the momentum function." Or roughly:

   Forcex = Mass x Acceleration.

   This relationship is not so much a natural law as a rule for assigning a magnitude to forces.

3. *Third Law:* "For every applied force there is an equal and oppositely directed reactive force." You push on the wall—the wall pushes back. The pusher and the pushed, the striker and the struck both experience forces of the same magnitude but of opposite direction.

IV. DYNAMICS: THE ANALYSIS OF FORCES AND THEIR EFFECTS

1. **Force.** A force in physics or engineering is something close to what we call "forces" in everyday life. Any push, pull, twist, etc., involves a force or forces. Forces are measured in physics according to their effects—according to Newton's Second Law, Force = Mass x Acceleration, or F= MA. Note that "mass" is that property of physical objects that causes them to resist changes in motion. "Inertia" is another term for mass.

2. **Weight.** Weight is a special force which results from the fact that a mass is being acted on by a gravitational field. A body on the surface of the earth has a certain weight because it has mass which, when acted on by the earth's gravitational field, would cause the body to be accelerated if it lost its support. If it could fall, its "weight" would result in its being accelerated toward the center of the earth.

3. **Torque.** A torque is a twisting force. It is a force that tends to induce rotary motion rather than straight line motion. Torque is typically measured in lb-ft. Thus, if we pull on a one-foot long wrench with a force of 100 lbs., we exert 100 ft-lb of torque on the nut. The same torque is generated by 50 lbs. on a two-foot wrench, etc.

4. **Friction.** Friction is a special kind of force produced by two bodies that are in contact. If a book is at rest on a desk and we try to push it, our efforts are resisted by what is known as static friction. Once we get it moving, if we stop pushing it, it comes to rest almost immediately. The retarding or stopping force is known as dynamic friction.

Rolling friction is the relatively low retarding force associated with the free rolling of objects, e.g. tires. Sliding friction is much greater than rolling friction and occurs whenever two objects in contact move with respect to one another without the benefit of any revolving elements. When a car is driven down the road without braking, it is being retarded by rolling friction (also by air drag, another type of force). When the wheels are locked, the vehicle is retarded by the sliding friction between the tires and on the road. In between these two cases, in situations with braking without wheel lock-up, the retarding forces are a complicated manifestation of the operation of the brakes, tires, and suspension system of the car. One measure of the sliding friction of cars is the coefficient of drag (Cd). This gives an indication of how hard it is to push the car along the road with its wheels locked. A typical Cd is 0.7. In this case, a force equal to 0.7 x the weight of the car is required to keep it sliding along the road. With a Cd and skid mark of known lengths, it is possible to estimate a vehicle's velocity before the start of the slide.

S(mph) = 5.5 x sqrt (Cd x length of skid)

5. **Momentum.** Momentum is the product of mass and velocity (momentum = mass x velocity). Momentum is a vector quantity as are velocity, acceleration and force. Momentum is conserved in impacts. That is, the sums of the various momentums of the bodies before the collision are the same as momentums after the collision. Thus, we can frequently compute the velocities of vehicles before a collision by knowing their speeds and directions of travel after the impact. Momentum is not the same as, and should not be confused with, energy.

6. **Work.** When a force acts on a body, work is said to be done on that body. The quantity of work, W, is given

by the formula: W = F x D, where "F" is the force that acts on the body and "D" is the distance through which it acts. Thus, a 100-lb. force acting for a distance of 10' (e.g., pushing an object against a resistance of 100# for 10') results in work = 1000 lb-ft being done on the object.

7. **Energy.** Energy is the capacity to do work. In Newtonian physics, the energy of a body is computed in two ways: either by computing its kinetic energy: KE = 1/2 M x V x V where "M" = mass and "V" = velocity; or by computing its potential energy with respect to a system of forces capable of doing work on the body. For example, the potential energy of a body in the earth's gravitational field is PE = W x H, where "W" is the weight of the body and "H" is its height above some reference point.

Computations involving kinetic energy are tricky but informative. For example, if a moving body crashes into a solid, non-yielding wall at 20 miles per hour, the kinetic energy dissipated in the crash = 1/2 M x 20 x 20 = 1/2 M x 400 = M x 200; but if it crashes into the wall at 40 miles per hour, the energy dissipated in the collision is 1/2 M x 40 x 40 = 1/2 M x 1600 = M x 800. Thus, four times as much energy is involved in the second crash as in the first (M x 800/M x 200 = 4). The speed has doubled but the energy has quadrupled!

8. **Power.** Power is the rate at which work is done. When work is done at the rate of 33,000 ft-lbs. per minute, one horsepower is produced. A "horsepower" is merely a conveniently sized unit for measuring power. Its connection with equine work capacity is tenuous. A human being can apparently work at a rate of about .35 horsepower for short periods of time, if all their skeletal muscles are being effectively used.

V. SPECIAL TERMS AND EXAMPLES OF NEWTONIAN ANALYSIS

1. **G's.** The term "g" forces or "g" loads is a convenient descriptive tool for technical discussions involving accelerations due to impact forces. One "g" is an acceleration equal to that generated by a free fall in the earth's gravitational field, i.e., 32.2 feet per second per second. Thus, a body acted on by a 0.5 g acceleration experiences a force equal to half its weight. This force acts in the direction of the acceleration.

   Frequently, during high speed collisions, accelerations in the range of 25 to 50 g's are generated. If a 150-lb. person is acted on by a 50-g retarding force during an accident, then a force of 7,500 lbs. acts on his body. Note that while a "g" value is really an acceleration, it is sometimes discussed as though it were a force. This is technically inaccurate but not harmful if it is clear what body the force acts on and if the mass of that body does not change during the application of the "g load."

2. **Moment.** The word "moment" in physics is really another word for torque. Generally, when we have two or more torques acting on a body, the result is said to produce a "moment" or net torque, which acts to rotate the body in a direction determined by the combined torque vectors.

3. **Pressure.** Pressure is force per unit area. Thus, if a 10-lb. weight has a contact area with another body of 1 square in., the pressure is 10 lb per square in. Normal atmospheric pressure at sea level is about 14.7 lb./sq.-in. (psi) This is the weight of a column of air 1 inch square extending up from sea level to the outer limit of the earth's atmosphere—roughly 80 miles high. A

barometer reading of about 30 inches of mercury (30 in hg) is about one atmosphere pressure or 14.7 psi. Thus, a 30-inch column of mercury weighs about as much as 80 miles of air since they both exert the same weight force per unit area.

4. **Stress.** Stress, like pressure, is also a force per unit area. However, while pressure acts on bodies, stress acts **in** a body. If you pull on both ends of a steel bar of 1 in. cross-sectional area with a force of 100 lbs., you'll generate a **tensile stress** inside the bar of 100 psi . If I pull across the bar so as to try to split it in half, I generate a **sheer stress**. Note that a mild steel bar would be able to absorb tensile stresses in excess of 50,000 psi without breaking so that my efforts to pull it apart result in only trivial stresses in the bar.

VI. ACCIDENT RECONSTRUCTION PROBLEMS

1. Suppose a 3000 lb. car were to crash into the rear of a truck that weighs 30,000#. The road is asphalt and the coefficient of drag is 0.7, the car is going 10 miles per hour, the truck is at rest.

    The momentum of the system before the collision is:

    > Momentum = Mass (car) x Velocity (car) + Mass (truck) x Velocity (truck)
    > = 3000 x 10 + 0 x 30,000
    > = 30000

    Since the external forces acting on the system during the impact are minimal (the brakes of the truck are off) momentum is conserved. Thus, after the accident the following relation holds true:

    > Momentum before =Momentum after

$$30{,}000 = 3000 \times V + 30{,}000 \times V$$
$$V = 30{,}000/33{,}000 = 0.9 \text{ mph}$$

where "V" is now the post impact velocity of the vehicles. (We assume here that the truck and the car are moving with the same velocity after impact.)

Thus, the car loses 9.1 mph = 13.3 ft/sec due to the action of the retarding impact force.

If the car is shortened about 8" in this impact, then the distance through which the retarding force acts is about 12" (the truck starts moving during the impact; assume it moves about 4") so that the car travels about 1' during the impact with an average velocity of about 8 ft/sec. Thus, the duration of the impact can be estimated as follows:

Distance = Velocity x Time
Time = Distance/Velocity
T = 1 ft/(8 ft/sec) = 0.125 sec

So the car decelerates from 14.7 ft/sec to 1.3 ft/sec in a time of 0.125 sec. Thus, its average deceleration is:

13.4 ft/sec / 0.125 sec = 107 ft/sec/sec = 3.33 g

The average force acting on the car then is 3.33 x 3,000# = 9990#

This is also the force that acts on the truck (Newton's Third Law: Every force has an equal and opposite reaction force): So that the average acceleration of the truck is: 30,000/9990 = 3 ft/sec/sec = 0.1 g

For Accident Statistics Motor Vehicle Fatal Crashes by Day of Week, Time of Day, and Weather and Light Conditions (percent) [Source: Bureau of National Transportation Statistics] go to:

http://www.car-accidents.org/stats-conditions.htm

Chapter 4

PATHOLOGY

Weather and Estimated Time of Death in Humans

One of the responsibilities of the forensic pathologist is to estimate the time of death and an accurate assessment is of great importance to police in narrowing down the list of suspects. It can allow police to pinpoint the time during which they need to find out what the suspects were doing and allows them to eliminate people who have an alibi for that period from there enquiries. Can you imagine the possible suspects on a timing window of 8 to 12 hours? What if we, as **meteorologists** could help reduce that window to, say 4 hours or less?

On a hot day in August in the Great Lakes, the body of an elderly man was found in a ditch next to a country road. He was laying supine (flat on his back), dressed in overalls, a short-sleeved work shirt and socks without shoes. His abdomen and legs were covered by a blanket. There was extensive maggot infestation of the head and neck, partially obscuring a scalp defect of the back of his head. In addition, there was a ligature (necklace, rope or something to show trauma) encircling the neck. The skin of the upper chest, neck and head was markedly darkened. The rest of the body was not decomposed. There was no apparent blood on the ground surrounding the decedent (dead person).

The scalp defect was a 4 < cm gaping laceration with no injury to the underlying bone or brain. The ligature was a small towel tightly compressing the neck and knotted in the back. There was no rigor mortis and livor mortis was present was present on the posterior aspect of the body. The stomach contained fragments of sausage, brown liquid and yellow-white semi-solid food particles. Approximately a dozen maggots were collected and preserved in alcohol. A diagnosis of ligature strangulation was rendered. The manner of death was ruled as homicide.

The next day the sheriff called the pathologist and asked about the time of death. The deputy was holding a man in custody that had been seen with the decedent two days prior to the body being found. The suspect was out of town on business the day before the victim's body was discovered. The suspect had good motive for the murder because of a soured business deal with the victim. Prior to making a formal arrest, the officer needed to make sure the postmortem interval (PMI) was consistent within two days. What should the pathologist tell the deputy about the PMI in this case?

One of the most frequently asked questions during a death investigation concerns the time of death. Unfortunately, determining the exact time of death from an examination of the body isn't possible. Numerous findings must be interpreted to give a reasonable estimate of the PMI. In this particular case, the time of death was a key element to the arrest because the suspect had an alibi for the day prior to the discovery of the body.

In this case, differential decomposition occurred because a head injury caused an open wound. Blood is an excellent stimulus for flies to lay eggs and for maggot activity which accelerated decompositional changes of the head. Since the rest of the body hadn't decomposed, an estimation of the PMI was made by evaluating the area of least decomposition. *Environmental temperature* is the most important factor in determining the rate of decompositional change after death. The decompositional changes in the head and the lack of rigor mortis could occur in 10–20 hours in temperatures circa (approx.) 90 degrees F. Therefore, in this case, the appearance of the body suggested a PMI of less than 24 hours.

The maggots appeared freshly hatched and were also consistent with being less than 30 hours of age. This information correlates with other findings since it is well established that flies lay eggs on a body very shortly after death. This would suggest the decedent died the day before his body was discovered. Therefore the "prime" suspect with the strong alibi does not appear to have committed the murder, and the investigators need to renew their search for the killer.

To a trained meteorologist who has an interest in this type of forensic science may look back and wonder how long the temperature was 90 degrees. The peak warming time of the day is usually noon to 4 p.m., and the best time to get close to 90 degrees. But in a 24 hour period, what would the hourly temperature be? Would the amount of cooling change the PMI? Would there be a measurable temperature change from the ground to two meters above the ground? What about absorption of the suns rays? Black asphalt vs. concrete will show diurnal variations, and during mid day when heating is the greatest, the concrete will be much cooler. This could change the PMI.

A difference in location yet the same time can have an impact of the decomposition of a body. Following death, the body begins to cool at a rate which depends on a range of factors. A naked body will cool faster than a clothed one; a large adult will cool slower than that of an infant; a body in a prone position will cool faster than a body slumped in the corner; a body exposed to air currents will cool faster than one in a protected area. Other factors such as topography will have an influence.

In 1978, a case occurred in a remote part of New South Wales which illustrates this rather dramatically. The local telephone operator suspected foul play when she could not connect a call to a remote farmhouse after a number of attempts. Eventually a young child answered. After some questioning the child explained that she did not '. . .like Mummy any more because she is turning black.' When the police arrived they found two bodies in quite different states of decay even though they had been killed at the same time.

The father had been shot on the kitchen and lay dead on the cool linoleum floor. The mother was found shot dead in bed and was in a far more advanced state of decomposition because the electric blanket had been left on. Because the body had been kept warm the bacteria found naturally in the human body had multiplied rapidly and begun the decay process. Flies had laid eggs in the body and maggots had hatched and grown. The state of the woman's corpse had deteriorated rapidly, while very little change occurred in the male victim.

Meteorologists, especially forensic meteorologists, have a really good idea on this concept and how to deal with the complexities of temperature deviations.

To have an idea how temperature works on the body, we have to know what the formulas are and how the data is entered in and learn the basics of decomposition over time.

(Referenced from the Medical Algorithms Project. Developed by the *Institute for Algorithmic Medicine,* a non-profit Texas Corporation.

Overview of formulas: The temperature of a body can be used to estimate the time of death.

Method 1

time since death in hours = ((rectal temperature at time of death in °F) - (rectal temperature in °F at time t)) / 1.5

Method 2

time since death in hours = ((rectal temperature at time of death in °C) - (rectal temperature in °C at time t)) + 3

Limitations:

- Body temperatures at the time of death that are above or below normal can give misleading results.

References:

Green MA Wright JC. Postmortem interval estimation from body temperature data only. Forensic Science International. 1985; 28: 35-46.

Knight B Nokes L. Chapter 2: Temperature-based methods I. pages 3–45 (pages 34–35). IN: Henssge C

Knight B et al. The Estimation of the Time Since Death in the Early Postmortem Period. Arnold Publishers. 1995.

Equation of De Saram

Overview: The temperature of a body can be used to estimate the time of death.

time since death in hours = ((((rectal temperature at time of death in °F) - (rectal temperature in °F at time t1)) / ((rectal temperature at time t1 in °F) - (rectal temperature in °F at time t2))) * (t2 - t1)) + 0.75

where:

- 0.75 hours (45 minutes) is a factor that can be added to take into account any delay in cooling

Limitations:

- Body temperatures at the time of death that are above or below normal can give misleading results.

References:

De Saram G Webster G Kathirgamatamby N. Post-mortem temperature and the time of death. J Crim Law Criminol Police Sci. 195; 1: 11–24.

Knight B Nokes L. Chapter 2: Temperature-based methods I. pages 3–45 (page 35). IN: Henssge C Knight B et al. The Estimation of the Time Since Death in the Early Postmortem Period. Arnold Publishers. 1995.

Method of Green and Wright

Overview: The method of Green and Wright for estimating the time of death uses rectal and environmental temperatures. It assumes that a double exponential mode adequately describes how the body cools.

gradient G = ((rectal temperature in °C at time 2) - (rectal temperature in °C at time 1)) / (time interval in hours between time 1 and time 2

where:

- The original equation shows temperature at time 1 minus temperature at time 2 but states that G is always negative since the latter temperature is always greater than the former.

gradient A = (((rectal temperature in °C at time 1) + (rectal temperature in °C at time 2)) / 2) - (average environmental temperature in °C)

reduced theta = ((rectal temperature at the time of death) - (((rectal temperature in °C at time 1) + (rectal temperature in °C at time 2)) / 2)) / ((rectal temperature in °C at time of death) - (average environmental temperature in °C))

where:

- The rectal temperature at the time of death is assumed to be 37°C.

Estimation of F

For reduced theta between 0 and 0.639 the rise in F is linear and is approximated by the equation derived in JMP:

F = (1.8826 * (reduced theta)) - 0.026636

For reduced theta from 0.639 to 0.96 the rise in F is exponential and can be approximated by the equation derived in JMP.

F = (317.62648 * ((reduced theta)^4)) - (926.6809 * ((reduced theta)^3)) + (1014.8223 * ((reduced theta)^2)) - (491.3695 * (reduced theta)) + 89.622192

Calculation of Time Since Death

time since death in hours = (-F) * ((gradient A) / (gradient G))

Limitations:

- The method assumes that the rectal temperature at the time of death was normal which may or may not be an accurate assumption.
- The method assumes that the environmental temperature remains relatively constant over the time since death.

References:

Green MA Wright JC. Postmortem interval estimation from body temperature data only. Forensic Science International. 1985; 28: 35–46.

Green MA Wright JC. The theoretical aspects of the time dependent Z equation as a means of postmortem interval estimation using body temperature data only. Forensic Science International. 1985; 28: 53–62.

Knight B Nokes L. Chapter 2: Temperature-based methods I. pages 3-45 (pages 22 and 38). IN: Henssge C Knight B et al. The Estimation of the Time Since Death in the Early Postmortem Period. Arnold Publishers. 1995.

Method of Fiddes and Patten

Overview: The method of Fiddes and Patten for estimating the time of death uses rectal and environmental temperatures. It uses the virtual cooling time which is the time it takes a body to reach a 15% difference between rectal and environmental temperatures.

Variables:

(1) rectal temperature in °C at the time of death (normally 37
(2) rectal temperature in °C at time 1
(3) rectal temperature in °C at time 2
(4) time interval in hours between time 1 and time 2
(5) environmental temperature in °C at time of death

percent fall in rectal temperature at time 1 = ((rectal temperature in °C at time of death) - (rectal temperature at time 1)) / ((rectal temperature in °C at time of death) - (environmental temperature in °C))

percent fall in rectal temperature at time 2 = ((rectal temperature in °C at time of death) - (rectal temperature at time 2)) / ((rectal temperature in °C at time of death) - (environmental temperature in °C))

Estimating Percent of Virtual Cooling Time from

Percent Fall in Temperature Difference

The percent fall in rectal temperature correlates with the percent fall in the virtual cooling time in a semilogarithmic manner (fall in temperature semilogarithmic percent virtual cooling time linear). Using data in Hennsge (1995 Figure 2.22 page 36) the fall in temperature difference and virtual cooling time can be approximated by equations from JMP:

For percent fall in temperature from 0 to 30 percent

percent of virtual cooling time = (0.00375 * ((percent fall in temperature) ^ 2)) + (0.5825 * (percent fall in temperature)) + 0.075

For percent fall in temperature from 30 to 60 percent
percent of virtual cooling time = (0.005 * ((percent fall in temperature) ^ 2)) + (0.49 * (percent fall in temperature)) + 1.7

For percent fall in temperature from 60 to 85 percent

percent of virtual cooling time = (0.0546231 * ((percent fall in temperature) ^ 2)) - (5.972111 * (percent fall in temperature)) + 211.30905

These are calculated for the percent fall in temperature seen at time 1 and time 2.

Estimating the Time Since Death

difference in percent virtual cooling time between time 1 and time 2 =
= (percent of virtual cooling time at time 2) - (percent of virtual cooling time at time 1)

100% of virtual cooling time in hours = (time interval in hours between time 1 and time 2) * 100 / (difference in percent virtual cooling time between time 1 and time 2)

time since death in hours = (100% of virtual cooling time in hours) * (percent fall in rectal temperature at time 1)

Limitations:

- The precise rectal temperature at the time of death may not be 37°C. The more that the actual rectal temperature deviates from this value then the more inaccurate the estimates.

- The environmental temperature needs to be relatively constant for the method to provide meaningful estimates.

References:

Knight B Nokes L. Chapter 2: Temperature-based methods I. pages 3–45 (pages 20–21 and 35–36). IN: Henssge C Knight B et al. The Estimation of the Time Since Death in the Early Postmortem Period. Arnold Publishers. 1995. (NOTE: It appears as if the legends for Figure 2.22 page 36 are switched. This would correlate with Figure 8 page 21 and would make more sense with the text.)

Method of Al-Alousi and Anderson

Overview: The method of Al-Alousi and Anderson to estimate the time of death uses organ and environmental temperature to calculate a temperature difference ratio which is then compared to a cooling curve. The method uses different cooling curves depending on whether the body is clothed or naked.

Organ temperature to measure:

(1) rectum
(2) liver
(3) brain
(4) any organ for which a cooling curve is available
(5) Thigh or other "meaty" part of the body where a steady temperature reading can be obtained.

temperature difference ratio = ((organ temperature in °C at time 1) - (environmental temperature in °C at time 1)) / ((organ temperature in °C at time of death) - (environmental temperature in °C at time 1))

Liver Body Naked

For temperature difference ratio >=0.23

mean time since death in hours at time 1 = (-39.477 * ((temperature difference ratio)^3)) + (104.578 * ((temperature difference ratio)^2)) - (103.96 * (temperature difference ratio)) + 38.859

For temperature difference ratio >=0.06

mean time since death in hours at time 1 = (863.727 * ((temperature difference ratio)^2)) - (405.879 * (temperature difference ratio)) + 67.726

For ratio < 0.06 time of death is assumed to be > 50 hours.

Liver Body Covered

For temperature difference ratio >=0.5

mean time since death in hours at time 1 = (-40 * (temperature difference ratio)) + 40

For temperature difference ratio >=0.17

mean time since death in hours at time 1 = (163.487 * ((temperature difference ratio)^2)) - (191.52* (temperature difference ratio)) + 75.143

For ratio < 0.17 time of death is assumed to be > 50 hours.

Rectum Body Naked

For temperature difference ratio >=0.22

mean time since death in hours at time 1 = (29.677 * ((temperature difference ratio)^2)) - (61.847 * (temperature difference ratio))+32.17

For temperature difference ratio >=0.01

mean time since death in hours at time 1 = (320.2587 * ((temperature difference ratio)^2)) - (205.8496 * (temperature difference ratio)) + 49.8496

For ratio < 0.01 time of death is assumed to be > 50 hours.

Rectum Body Covered

For temperature difference ratio >=0.33

mean time since death in hours at time 1 = (27.4789 * ((temperature difference ratio)^2))-(66.3977 * (temperature difference ratio)) + 38.91878

For temperature difference ratio >=0.01

mean time since death in hours at time 1 = (211.783 * ((temperature difference ratio)^2)) - (159.305 * (temperature difference ratio)) + 49.59259

For ratio < 0.01 time of death is assumed to be > 50 hours.

Limitations:

- The body temperature at the time of death may not be 37 °C.

- If the environmental temperature varies widely and is not constant then the environmental temperature used may be misleading.
- Above 40–50 hours the curves flatten and may not be accurate.
- The cooling curves show a hatched area +/- 1 standard deviation so that the estimates are only rough time indicators.

References:

Knight B Nokes L. Chapter 2: Temperature-based methods I. pages 3–45 (pages 23–24 and 38). IN: Henssge C Knight B et al. The Estimation of the Time Since Death in the Early Postmortem Period. Arnold Publishers. 1995. (NOTE: The time designations in the legend for the brain curves to Figure 2.10 page 23 may not be right; check the original reference).

A new computer program is currently being tested which takes all of these formulas into account along with non-linear algorithms to create a more realistic regression of temperature over time and introduces other facets of forensic evidence to help narrow down the estimated time of death (ETOD).

Unlike the versions that are currently running where it's derived by body temperature alone. This new program takes into account other factors in the PMI.

Timely Death

The timing of death is both an art and a science. Unless the death is witnessed, it is impossible to determine the exact time of death. The Medical Examiner (ME) can only "estimate" the approximate time of death and it is important to note that this "estimated time of death" can vary greatly from the "legal" time of death, which is the time recorded on the death certificate, or the "physiologic" time of death, which is when vital functions actually cease. The "legal" time of death is the time the body was discovered or the time a doctor or other qualified person pronounced the victim dead. These "times of death" may differ by days, weeks, even months, if the body is not found until well after "physiologic" death has occurred.

For example, if a serial killer kills a victim in July, but the body is not discovered until October, the "physiologic" death took place in July, but the "legal" death is marked as October. The ME's "estimated time of death" would be July. That said, the ME can estimate the "physiologic" time of death with some degree of accuracy. He uses the decomposition changes that occur in the human body after death to help him in this endeavor. These changes consist of measuring the drop in body temperature, the degree of rigidity (rigor mortis), the degree of discoloration (livor mortis or lividity), the stage of body decomposition, stomach contents, and other factors. Bodies found in water present special problems in this regard.

A project study is underway to determine if it is feasible for EMS personnel to assist in determining the time of death in areas where it's physically impossible for the medical examiner to attend the crime scene. Protocol standards will have to be written and approved by the local prosecuting attorney and fall in line with the National Guidelines of Death investigations. This would be a benefit for those areas that are not fully staffed with enough medical examiners to investigate every crime scene due to distance or other limiting factors.

Body Temperature:

> Normal body temperature during life is 98.6 degrees F. This will vary if a body was attacked or there was physical activity prior to death, then the body temperature would be higher (in some cases as much as 2–4 degrees). After death, the body loses heat progressively until it equilibrates with that of the surrounding medium. The rate of this heat loss is approximately 1.5 degrees per hour until the environmental temperature is attained, then it remains stable. Obviously, this measure is greatly affected by location and clothing. A body in the snow in Michigan in January and one in a Louisiana swamp in August will lose heat at widely divergent rates. These factors must be considered in any "estimate" of time of death.

The criminalist who processes the scene should take a body temperature and measure the temperature of the surrounding medium—air, water, snow, earth (if the body is buried). Ideally, the body temperature is taken rectally or through the liver. Obviously, the sooner after death the body is found, the more accurately time of death can be assessed by this method. Once the body reaches ambient temperature, all bets are off. Air temperature at 2 meters isn't sufficient because the body is on the ground, so take temperatures that are relative to the surroundings. Keep in mind that daytime temperatures where the sun shines will differ to shaded readings (such as the standard readings obtained by official standards).

Rigor Mortis:

Rigor mortis is the stiffening and contraction of the muscles due to chemical reactions that take place in the muscle cells after death. It typically follows a predictable pattern. Rigidity begins in the small muscles of the face (cheek area) and neck and progresses downward in a "head-to-toe" fashion to the larger muscles. The entire process takes about 8–12 hours. At that time, the body is completely stiff and is "fixed" in the position of death. Then, the process reverses itself, with rigidity being lost in the same fashion, beginning with the small muscles and progressing to the larger ones. This process begins 18 to 36 hours after death and is usually complete within 48 hours. So, rigor is only useful in the first 48 to 60 hours after death. Physical activity will speed up the process.

(Rigor Mortis (Photo Courtesy of Jay Dix and Michael Graham— Time of Death, Decomposition and Identification, an Atlas.)

The reason for the rigidity is the loss of adenosine triphosphate, or ATP, from the muscles. ATP is the compound that serves as energy for muscular activity and its presence and stability depend upon a steady supply of oxygen and nutrients, which are lost with the cessation of cardiac activity. The later loss of rigidity and the appearance of flaccidity (relaxation) of the muscles occur when the muscle tissue itself begins to decompose. Rigor is one of the least reliable methods for determining time of death because it is extremely variable. Heat quickens the process, while cold slows it. Obese people may not develop rigor, while in thin victims it tends to occur rapidly. If the victim struggled before death and consumed much of his muscular ATP, the process is hastened. Exercise and heat also play a roll in speed of rigor. After a 24 mile marathon, the body could be fully rigor within six hours because it can't replace the oxygen that was lost during exercise.

Simplified table of Rigor Mortis

Temperature of body	Stiffness of body	Time since death
Warm	Not stiff	Not dead more than three hours
Warm	Stiff	Dead between 3 to 8 hours
Cold	Stiff	Dead between 8 to 36 hours
Cold	Not stiff	Dead in more than 36 hours

Lividity:

Lividity is caused by stagnation of blood in the vessels. It lends a purplish color to the tissues. The blood, following the dictates of gravity, seeps into the dependent parts of the body—along the back and buttocks of a victim who is supine after death. Initially, this discoloration can be "shifted" by rolling the body to a different position, but by six to eight hours, it becomes "fixed." If a body is found face down, but with fixed lividity along the back, then the body was moved at least six hours after death, but not earlier or the lividity would have "shifted" to the newly dependent area.

Lividity (Photo Courtesy of Jay Dix and Michael Graham—Time of Death, Decomposition and Identification, an Atlas.)

Body Decomposition:

At death, the body begins to decompose. Bacteria go to work on the tissues and by 24 to 36 hours the smell of rotting flesh appears and the skin takes on a progressive greenish-red color. By three days, gas forms in the body cavities and beneath the skin, and may leak fluid and split. From there, it gets worse. Add to this, predation by animals and insects and the body can become completely skeletonized before long. In hot, humid climes, this can happen in three or four weeks.

Stomach Contents:

The ME can often use the contents of the victims stomach to help determine time of death. After a meal, the stomach empties itself in approximately 4 to 6 hours, depending on the type and amount of food ingested. If a victim stomach contains largely undigested food material, then the death likely occurred within an hour or two of the meal. If the stomach is empty, the death likely occurred more than six hours after eating. Additionally, if the small intestine is also empty, death probably occurred some 12 hours or more after the last meal.

This table describes the process in the decomposition of the pig.

Stage	Description
Initial Decay	The cadaver appears fresh externally but is decomposing internally due to the activities of bacteria, protozoa and nematodes present in the animal before death

Putrefaction	The cadaver is swollen by gas produces internally, accompanied by odor of decaying flesh
Black putrefaction	Flesh of creamy consistency with exposed parts black. Body collapses as gases escapes. Odor of decay very strong
Butyric fermentation	Cadaver drying out. Some flesh remains at first, and cheesy odor develops. Ventral surface moldy from fermentation
Dry decay	Cadaver almost dry; slow rate of decay

Floaters:

"Floaters" are corpses found floating in a body of water. They present special problems for the ME in determining the time of death. Water temperature of course has an effect as do local tides and predators. The general rule regarding decomposition is that one week on dry land equals two weeks for a submerged body.

To become a "floater," a body must to be in the water long enough for tissue decomposition from bacteria to begin. This process forms gas as a byproduct, which collects beneath the skin and in body cavities. Bodies tend to sink, then rise again in several days when the gas forms, adding buoyancy. They thus become "floaters."

Under these circumstances, the hands and feet swell (several days), the outer layer of skin separates from the underlying tissues (five-six days), the skin of the hands and the nails separate (eight-ten days), and entire body swells shortly thereafter. Tissues become extremely fragile and are easily damaged during removal from the water. Timing of the "floating" depends upon several factors, including water temperature, currents, the size of the victim, and other variables. For example, a body will "float" after eight-ten days in warm water and two-three weeks if in colder water. Cold slows the process of decomposition.

As you can see, the timing of a victim's death is a very inexact science and is greatly altered by the environment. In cold areas, body temperature changes are magnified, but decomposition changes are slowed, the inverse is true for hot, humid climes. Add to this, predation by insects and animals and the ME's job can become difficult. (DPLyle,MD 2000).

However, with the use of a forensic meteorologist (FM), the timing error can be greatly reduced offering the investigators a smaller timing error or more information to help close the investigation

The same theory holds true for entomologist (study of insects). Tests have been done to determine rate of growth from egg to adult fly. However, these tests were performed at a **constant** temperature. Since insects are often found outside, how often do we observe a constant temperature for 12 hours? How about 24, 36 or 48 hours? It can't happen! Temperatures change and so does the growth rate of insects. Thus, a timing error is often incorporated due to the inexact science of fly-rearing. Is there a way to reduce the timing error? I believe so and it would take the use of sample data to compare against a non-linear differential equation to find the answer.

Now let's apply what we have learned using temperature to determine time of death.

Environmental Forensics and its Effects on Investigations

You are the death investigator at a scene where a 47-year old man has been found dead on the floor of his bedroom. You arrive at midnight, and immediately obtain a core rectal temperature of 92.4° F. Room temperature is 70° F. Rigor is present in the jaw, arms, hands, and feet. Livor is well-established but is not fixed. The man is dressed in his pajamas, and his height and weight as stated on his drivers' license are 6'3" and 195 pounds. These are consistent with his appearance.

1. Using the rough calculation where the body cools at 1.5 degrees per hour, what do you estimate the time of death to be?
2. You learn that the thermostat in the house was programmed to change at 11 PM from 76° F to 70° F. Would this change your estimate of time of death using method #1?
3. You learn that the thermostat in the house was programmed to change at 11 PM from 76° F to 70° F. Would this change your estimate of time of death using method #2?

If you noticed, the formulas asked for a single temperature. What if the temperature changed over time? Would that have an influence in the calculation? Back in the beginning of this book I had some numbers from my thermostat;

Main floor thermostat is located on main floor (5 feet high)

Second story bed	70.7
Main floor wall	69.7
Placed on carpet /main floor	67.9
Basement floor	65.7

There's a five degree difference between the basement floor and the second story bed. Would this five degree difference change the ETOD?

Can you imagine the offset if we used temperatures outside? It's imperative that we give the best possible dataset to aid investigators.

Eyes

Within minutes the cornea films over, and the white of the eye goes grey.

After around two hours the cornea goes cloudy, and within a day or two it goes opaque.

On the third day the gas makes the eyes bulge.
With advanced decomposition, the eyes retract.

Food in the stomach

(Found at autopsy)

A light meal is out of the stomach within 1 ½–2 hours

A medium meal is out of the stomach within 3–4 hours

A heavy meal is out if the stomach within 4–6 hours

There are variations: Liquid is digested faster than semi-solid food, which is digested fasted than solid food. Emotional state may also influence the rate of stomach emptying.

Vitreous Potassium

There is potassium in the body's intercellular fluid - much more than there is in the plasma, on the other side of the cell membrane. After death it starts to leak out so there is the same amount on either side of the membrane. This happens at a nice steady rate, allowing time of death to be established. Samples are normally taken from the vitreous fluid of the eye.

Insects

- Body Lice: they usually outlive their host by three—six days.
- Various insects: They like to lay their eggs on very fresh corpses. The eggs hatch out within eight-fourteen hours. After another eight-fourteen hours it sheds its skin and emerges as a bigger larva. This process is repeated several times, taking ten-twelve days in total. People who know about insects can therefore look at the larva, see which stage it is at, and work out time of death.

Samples should be taken and preserved and given to an entomologist.

Plants

- Grass/plants beneath an object wilt, turn yellow or brown and dies. The rate depends on type of plant, season, climate, etc.
- Seasonal plants or remnants may help indicate a range of time.

Samples should be collected and shown to a botanist.

Putrefaction

Purification begins after about two days. The process is faster in damp places or when the body is exposed to air. Decay is about eight times faster in the air than underground. Too cold or too hot and the process won't happen. In very hot temperatures the body will dry out and mummify instead.

People with a lot of fat will decay faster. People who died of bacterial disease will also decay faster. However, some poisons preserve the body.

2–3 days: green staining begins on the right side of the abdomen. Body begins to swell.
3–4 days: staining spreads. Veins go "marbled" - a brown-black discoloration
5–6 days: abdomen swells with gas. Skin blisters
2 weeks: abdomen very tight and swollen.
3 weeks: tissue softens. Organs and cavities start bursting. Nails fall off.
4 weeks: soft tissues begin to liquefy. Face becoming unrecognizable
4–6 months: formation of adipocerous, if in damp place. This is when the fat goes all hard and waxy.

Anamnestic Evidence:

This means evidence taken from the victim's daily habits. For example, three days' uncollected newspapers would suggest he has been dead three days. If he missed an appointment on a particular day then he was probably dead then. If all his food in the refrigerator is rotten and horrid it suggests he's been dead a while.

Chapter 5

FORENSIC ENTOMOLOGY ENVIRONMENTAL IMPACTS IN ESTIMATING TIME OF DEATH

Forensic (or medico-legal) entomology is the study of the insects associated with a human corpse in an effort to determine elapsed time since death. Insect evidence may also show that the body has been moved to a second site after death, or that the body has been disturbed at some time, either by animals, or by the killer returning to the scene of the crime. However, the primary purpose of forensic entomology today is to determine elapsed time since death.

Forensic entomology was first reported to have been used in 13th Century China and was used sporadically in the 19th Century and the early part of the 20th Century, playing a part in some very major cases. However, in the last 15 years, forensic entomology has become more and more common in police investigations. Most cases that involve a forensic entomologist are 72 hours or more old, as up until this time, other forensic methods are equally or more accurate than the insect evidence. However, after three days, insect evidence is often the most accurate and sometimes the only method of determining elapsed time since death.

After the initial decay, and the body begins to smell, different types of insects are attracted to the dead body. The insects that usually arrives first is the diptera, in particular the blow flies or calliphoridae and the flesh flies or sarcophagidae. The females will lay their eggs on the body, especially around the natural orifices such as the nose, eyes, ears, anus, penis and, depending on the type of assault, the vagina. If the body has wounds the eggs are also laid in such. Flesh flies do not lay eggs, but deposits larvae instead.

After some short time, depending on species, the egg hatches into small larvae. This larvae lives on the dead tissue and grows fast. After a little time the larva molts, and reaches the second larval instar. Then it eats very much, and it molts to its third instar. When the larvae are fully grown it becomes restless and begins to wander. It is now in its prepupal stage. The prepupae then molts into a pupae, but keeps the third larval instars skin, which becomes the so-called puparium. Typically it takes between one week and two weeks from the egg to the pupae stage. The exact time depends on the species and the temperature in the surroundings. A table of life histories to some species of blow flies and flesh flies:

Some development data on different species of blowflies (Calliphoridae) and fleshflies (Sarcophagidae)

Life histories of 11 species of blowflies and fleshflies reared at 27 degrees Celsius, and 50 percent relative humidity.

	No. of Gen.	Egg (Hrs)	First instar (Hrs)	Second instar (Hrs)	Third instar (Hrs)	Prepupa (Hrs)	Pupa (Days)	Total Immature (Days)
Sarcophaga cooley	29	–	24	18	48	96	9	16
Sarcophaga sherma	28	–	22	16	48	104	8	14
Sarcophaga bullat	18	–	26	18	54	112	12	17
Phormia regina	23	16	18	11	36	84	6	11
Protophormia terr	27	15	17	11	34	80	6	11
Lucilia sericata	29	18	20	12	40	90	7	12
Eucalliphora lila	27	22	22	14	36	92	6	13
Cynomyopsis cadav	17	19	20	16	72	96	9	18
Calliphora vomit	5	26	24	48	60	360	14	23
Calliphora vicina	5	24	24	20	48	128	11	18
Calliphora terran	4	25	28	22	44	144	12	20

(After Kamal, 1958)

An illustration of the blowfly life cycle:

8.3
SIZE/TIMELINE OF HYPOTHETICAL IMMATURE BLOW FLY DEVELOPMENT

    The theory behind estimating time of death, or rather the post mortem interval (PMI for short) with the help of insects are very simple: since insects arrive on the body soon after death, estimating the age of the insects will also lead to an estimation of the time of death. As you review the developmental stages, you'll see the temperatures that are fixed (not changing over time). We as meteorologist can start to see the benefit to our expertise and how it can change the developmental stage(s) of the insects.

How to estimate age of blowfly eggs, larvae, pupae and adults

Eggs:

When blow flies oviposit, their eggs has come very short in their development. The eggs are approximate two mm in length. During the first eight hours or so there are little signs of development. After that, and one can see the larvae through the chorion of the egg at the end of the egg stage. The egg stage typically lasts a day or so.

Larvae:

The blowfly has three instars of larvae. The first instar is approximately five mm long after 1.8 days, The second instar is approximately ten mm long after 2.5 days, The third instar is approximately 17 mm long after 4–5 days. Identifying the right instar is the easiest part, and is done relatively easy based on size of larvae, the size of the larva's mouth parts and morphology of the posterior spiracles. The time it takes to reach the different instars depends very much on microclimate, i.e. temperature and humidity.

Prepupae:

At the end of the third instar the larva becomes restless and starts to move away from the body. The crop will gradually be emptied for blood, and the fat body will gradually obscure the internal features of the larvae. We say that the larva has become a prepupa. The prepupa is about 12 mm long, and is seen 8–12 days after oviposition.

Pupa:

The prepupa gradually becomes a pupa, which darkens with age. The pupa which are about 9 mm in length are seen 18–24 days after oviposition. The presence of empty puparia should therefore tell the forensic entomologist that the person in question has been dead in more than approximately 20 days. Identification can be done based on the remaining mouth parts of the third instar larvae.

A more precise way to determine age of larvae and eggs is the use of rearing. For example: the body is found with masses of eggs on it, none have hatched. How long time is it since the eggs was oviposited? Note the time of the discovery, note the time when the first 1. Instar larvae occur. Subtract the first occurrence time with the discovery time, call this time A. Rear the blow flies to adults, let them mate, let them lay eggs on raw beef liver under conditions similar to the crime scene, take the time from oviposition to the first occurrence of 1. Instar larvae. Call this time B. By subtracting B-A, one gets C, which is an estimate of the time since oviposition to discovery. Similar calculations can be done for other instars as well. If one has good base-line data from before under different temperatures and for different species, one only needs to rear the flies to a stage where they can be identified, and that is the third stage or the adult stage.

One important biological phenomenon that occurs on cadavers is a succession of organisms that thrive on the different parts. i.e. beetles that specialize on bone will have to wait until bone is exposed. Predatory rove beetles or parasites that feed on maggots will have to wait until the blow flies arrive and lay their eggs.

The succession on cadavers happens in a fairly predictable sequence and can be used in estimating time of death if the body has been lying around for some time.

Table of succession on guinea pigs from Bornemizza 1957

There are several things to note about this table:

The first group to arrive is blow flies, followed shortly by staphylinids. As putrefaction develops, more groups arrive at the scene, with most groups present just before the body is drying out due to seepage of liquids. After the body is drying out, dermestids, tineids and certain mites will be the dominant animal groups on the body, and blow flies will gradually vanish. Note also how the fauna changes in the soil. This can also be used to estimate time since death.

Succession data can be incorporated in a database, and when the forensic entomologist investigates a case, he can use the taxa found on the body as input, and get an estimate of the time of death as output.

Several insects are specialized in living in very decayed dead bodies. One example is the cheese skipper, *Piophila casei*, where the larvae usually occur 3–6 months after death. The cheese skipper is a well known pest of cheese and bacon worldwide, and has a cosmopolitan distribution. Adult cheese skippers may occur early after death, but larvae occurs later.

In temperate regions dead bodies often appear in spring, after the snow is gone. The forensic entomologist and the forensic pathologist must then try to determine whether the death occurred during the winter or before the snow set in. If the death occurred before November, it is possible to find dead insects in and on the body. By analyzing the dead insect fauna and estimating when the insects probably died (this can be found by looking at meteorological records). Another hint is when the different adults stop flying before the winter. For example: There are some cases where the bodies have been found in the spring. In one case there were dead third stage blow fly larvae in the back of the mouth. The blow fly larva was of a species that is flying from May to October. It was from this conclusion that the eggs probably was laid during October, and since it was relatively few larvae, probably late in October. In another case, several live insects were found on a dead body, and also many dead third stage larvae. The dead larvae were found on the stomach, the arms, the shoulders, and inside the head.

A conclusion that the live insects had colonized the dead body in the spring and that the dead larvae had died during the winter. Based on the widespread occurrence of the larvae, it was likely that the body was colonized before October, probably in September.

If the death occurred in the winter things become difficult in outdoor settings, as very few insects are active in the winter. It is reported that larvae of the winter gnat, *Trichocera sp.* can develop on carrion in the winter. By estimating the age of these larvae, if present, it could be possible to estimate the PMI.

Although the amount of information discussed is well past the expertise of the meteorologist, it does give them a better idea as to the concept, application and effect the environment has on estimating time of death. If you are interested in knowing more about the collection and preservation of fauna, keep reading.

There are two main ways of using insects to determine elapsed time since death:

Using successional waves of insects

Using maggot age and development.

The method used is determined by the circumstances of each case. In general, the first method is used when the corpse has been dead for between a month up to a year or more, and the second method is used when death occurred less than a month prior to discovery.

The first method is based on the fact that a human body, or any kind of carrion, supports a very rapidly changing ecosystem going from the fresh state to dry bones in a matter of weeks or months depending on geographic region. During this decomposition, the remains go through rapid physical, biological and chemical changes, and different stages of the decomposition are attractive to different species of insects. Certain species of insects are often the first witnesses to a crime. They usually arrive within 24 hours of death if the season is suitable *i.e.* spring, summer or fall in Canada and can arrive within minutes in the presence of blood or other body fluids.

These first groups of insects are the calliphoridae or blowflies and the muscidae or houseflies. Other species are not interested in the corpse when the body is fresh, but are only attracted to the corpse later such as the piophilidae or cheese skippers which arrive later, during protein fermentation. Some insects are not attracted by the body directly, but arrive to feed on the other insects at the scene. Many species are involved at each decomposition stage and each group of insects overlaps the ones adjacent to it somewhat. Therefore, with knowledge of the regional insect fauna and times of carrion colonization, the insect assemblage associated with the remains can be analyzed to determine a window of time in which death took place. This method is used when the decedent has been dead from a few weeks up to a year, or in some cases several years after death, with the estimated window of time broadening as time since death increases. It can also be used to indicate the season of death *e.g.* early summer. Knowledge of insect succession is required for this method to be successful.

The second method, that of using maggot age and development can give a date of death accurate to a day or less, or a range of days, and is used in the first few weeks after death. Maggots are larvae or immature stages of diptera or two-winged flies. The insects used in this method are those that arrive first on the corpse, that is, the calliphoridae or blowflies. These flies are attracted to a corpse very soon after death. They lay their eggs on the corpse, usually in a wound, if present, or if not, then in any of the natural orifices. Their development follows a set, predictable, cycle.

The insect egg is laid in batches on the corpse and hatches, after a set period of time, into a first instar (or stage) larva. The larva feeds on the corpse and moults into a second instar larva. The larva continues to feed and develop into a third instar larva. The stage can be determined by size and the number of spiracles (breathing holes). When in the third instar, the larva continues to feed for a while then it stops feeding and wanders away from the corpse, either into the clothes or the soil, to find a safe place to pupate. This non-feeding wandering stage is called a prepupa. The larva then loosens itself from its outer skin, but remains inside. This outer shell hardens, or tans, into a hard protective outer shell, or pupal case, which shields the insect as it metamorphoses into an adult. Freshly formed pupae are pale in color, but darken to a deep brown in a few hours. After a number of days, an adult fly will emerge from the pupa and the cycle will begin again. When the adult has emerged, the empty pupal case is left behind as evidence that a fly developed and emerged.

Fig. 3 Isomorphen diagram for *C. albiceps*, showing all stages from oviposition to eclosion. Areas between lines represent identical morphological stages at various temperatures. Each line represents identical morphological changes of this holometabolous insect: for values (mean days ± SD) see Table 1

Each of these developmental stages takes a set, known time. This time period is based on the availability of food and the temperature. In the case of a human corpse, food availability is not usually a limiting factor.

Insects are 'cold blooded', so their development is extremely temperature dependent. Their metabolic rate is increased with increased temperature, which results in a faster rate of development, so that the duration of development decreases in a linear manner with increased temperature, and *vice-versa*.

An analysis of the oldest stage of insect on the corpse and the temperature of the region in which the body was discovered leads to a day or range of days in which the first insects oviposited or laid eggs on the corpse. This, in turn, leads to a day, or range of days, during which death occurred. For example, if the oldest insects are 7 days old, then the decedent has been dead for at least 7 days. This method can be used until the first adults begin to emerge, after which it is not possible to determine which generation is present. Therefore, after a single blowfly generation has been completed, the time of death is determined using the first method, that of insect succession.

PROCEDURE

The first and most important stage of the procedure involved in forensic entomology involves careful and accurate collection of insect evidence at the scene. This involves knowledge of the insect's behavior; therefore it is best performed by an entomologist. Unfortunately, the entomologist is often not called until after the body has been removed from the death site. They usually see the remains at the morgue, and in some cases, do not actually see the remains at all, so their evidence is **dependent** on **accurate** collection by the investigating officers.

Environmental Forensics and its Effects on Investigations

Fig. 2. Isomegalen-diagram for *L. sericata* larvae from hatching to peakfeeding. Time is plotted against temperature, each line representing identical larval length (mm). The small graph shows the analogous egg-periods between 15 and 40 C.

Fig. 3. Isomorphen-diagram for *L. sericata*, showing all stages from oviposition to eclosion (15–34 C). Areas between lines represent identical morphological stages at various temperatures. Where a = egg; b = 1st instar; c = 2nd instar; d = 3rd instar; e = postfeeding larva (i.e. prepupa); f = pupa; g = imago. Each line represents identical morphological changes of this holometabolous insect.

Table 9.1 Various Forensic Fly (Calliphoridae, Phoridae, and Sarcophagidae) Species with Their Corresponding Minimum Developmental Thresholds and Experimentally Determined Centigrade Degree-Day (CDD) and Centigrade Degree-Hour (CDH) Accumulations for Egg-to-Adult Emergence, Calculated from Data in Six Sources

Species	Dev. Min. (°C)	Dev. Environ. (°C)	Total Dev. Time (h)	CDD	CDH	Data Ref. in Source
Source: Kamal (1958)[a]						
Calliphora terraenovae	6	26.7	551	475	11406	Table 1
Calliphora vicina	6	26.7	508	438	10516	Table 1
Calliphora vomitoria	6	26.7	854	737	17678	Table 1
Cynomyopsis cadaverina	6	26.7	439	379	9087	Table 1
Eucalliphora lilaea	10	26.7	330	230	5511	Table 1
Phaenicia sericata	10	26.7	348	242	5812	Table 1
Phormia regina	10	26.7	309	215	5160	Table 1
Protophormia terraenovae	10	26.7	301	209	5027	Table 1
Sarcophaga bullata	10	26.7	498	347	8317	Table 1
Sarcophaga cooleyi	10	26.7	402	280	6713	Table 1
Sarcophaga shermani	10	26.7	382	266	6379	Table 1
Source: Greenberg (1991)[b]						
Calliphora vicina	6	10	1647	275	6588	Table 3
	6	12.5	1069	290	6949	Table 3
	6	19	583	316	7579	Table 3
	6	23	460	364	8740	Table 3
	6	10	1676	279	6704	Table 7[d]
	6	12.5	1063	288	6910	Table 7[d]
	6	19	562	304	7306	Table 7[d]
	6	23	446	297	7136	Table 7[d]
			Mean	302	7239	

COLLECTING, PRESERVING AND PACKAGING SPECIMENS

Collection

Samples of insects of all stages should be collected from different areas of the body, from the clothing and from the soil/carpet *etc*. Insects will often congregate in wounds and in and around natural orifices.

The two main insect groups on bodies are flies (diptera) and beetles (coleoptera).

Both types of insect look very different at different stages of their lives.

Flies can be found as:-
eggs (in egg masses usually)
larvae or maggots (in a range of sizes from 1–2 mm to 17 mm)
pupae and/or empty pupal cases
adults.

Eggs - are very tiny, but are usually laid in clumps or masses, and are usually found in a wound or natural orifice, but may be found on clothing *etc*. They can be collected with a child's paint brush dipped in water or with forceps. Half should be preserved in 75% alcohol or 50% isopropyl alcohol. The rest should be placed in a vial with a little damp tissue paper to prevent dehydration. If it will be more than a few hours before the entomologist receives them, they should also be given a small piece of beef liver. Make sure there is tissue or sawdust present if liver is added, to prevent drowning. They need some air. Newly emerged maggots can escape through holes, so a paper towel held over the top of the vial with a rubber band is excellent, as long as the vial stays upright! (No lid other than the paper towel is needed.)

Maggots - collect a range of sizes. Maggots will be found crawling on or near the remains and may be in **maggot masses**. The masses generate a lot of heat, which speeds up development.

Therefore, please note:

1. The site of maggot masses
2. The temperature of each mass (thermometers can be acquired cheaply at drugstores) or if no thermometer available, please estimate size of mass.
3. Label which maggots come from a particular mass.

Large maggots are usually older so are most important, but smaller maggots may belong to a different species so both large and smaller maggots should be collected, with the emphasis on larger maggots. Collect samples of maggots from different areas of the body and the surrounding area, and keep them separate.

As mentioned before, third instar larvae leave the food source to find a suitable area in which to pupate. They may wander some distance from the body so the soil for a meter or two around the body should be carefully sifted. Some may burrow down into leaf litter, so the soil below the corpse should be checked for several centimeters. If the remains were on a slope, the body fluids will seep downhill and insects will be found here, feeding on the fluids. This means that a very intensive search of the corpse, the clothes and the surrounding area must be made in order to get the entire picture.

When collected, a proportion of the larvae should be preserved immediately for two reasons. Firstly, to show the entomologist, if s/he is not present at the scene, what stage the larvae were when collected, as if they are then placed on meat, they will continue to develop, giving a misleading impression to the entomologist when they are examined. Secondly, to produce as evidence in court, if there are lots of maggots on the body, preserve approximately half of all sizes. If there are only 20–30, preserve 1 or 2. Preserve the specimens by immersing them in hot water for a few minutes, then putting them in 70% alcohol or 50% isopropyl alcohol. If no hot water available, put straight in preservative. Don't forget that most should be kept alive. A sample should contain about 100 maggots (of each size if possible).

The living specimens should be placed in a vial, with air and food, as for the eggs. There should be only enough maggots to cover the bottom of the vial. Too many in one vial will drown.

Pupae and Empty Pupal Cases—these are extremely important and are easy to miss. They are often found in clothing, hair or soil near the body. Pupae like dry, secure areas away from the wet food source in which to pupate so pockets, seams and cuffs are likely hiding places. If the remains are found indoors, they may have traveled some distance and be under clothing, rugs, boxes *etc*. They range from 2–20 mm, and are oval, like a football. They are dark brown when completely tanned. An empty pupal case is very similar but is open at one end, where the adult fly has emerged. They need some air, so secure a paper towel over vial as for eggs, as although the pupae are immobile, if they emerge during transit, an adult can get out of anything! A piece of tissue in the vial will help to avoid breakage as they are quite vulnerable. This can be *very slightly* moistened with water, but be careful not to drown them. The moisture isn't necessary if the journey is not long. **Do not preserve pupae!** They won't grow, so the reasons for preserving larvae do not apply, and it is almost impossible to identify a pupa until it emerges as an adult. I also cannot determine its exact age until I find out the day on which it emerges. If a pupa is found when a pale color, it is just entering pupation, so please keep that specimen separate and label as pale colored, as it will darken in a few hours. Such a specimen can be aged to a matter of hours.

Adult Flies - are less important. They are only of use in indicating which species of insect are likely to develop from the corpse, as you cannot determine whether an adult has developed on the corpse, or has just arrived from somewhere else to oviposit, unless it emerged only an hour or so earlier. **If an adult has crumpled wings**, it may have just emerged, so IS still important as it can be linked to the body. It should be collected, labeled as such, and kept separate. Adults can be collected by net or by using an inverted vial. They can be left in the vial without air or food, as I will kill them as soon as I receive them. The presence of empty pupal cases, however, indicates that an insect has developed on the corpse and reached adulthood. This can be very important as it indicates that at least one generation of flies of this species has completed development on the corpse.

Beetles - can be found as adults, larvae or grubs, pupae and also as cast skins. All stages are equally important. They move fast and are often found under the body, and in and under clothing. They can be placed in vials with some air. They only need to be fed if it will be more than 24h before they reach an entomologist. If necessary they can be fed extra maggots. They are cannibals so should not be placed in the same vial!!

Other Insects - other insects may be present. If you are not sure whether it's an insect, collect it anyway and place in a vial.

Other Samples - Soil and leaf litter samples will also be useful. About a coffee can size of **soil** from under of very near the body is useful. If the soil below the body is extremely wet, it is better to collect the soil from near the remains.

Labeling - Insects collected from one part of the body should be kept separate from those from another area. Different species should be kept separate as beetle larvae feed on fly larvae! If they look different, separate them. Each vial should be labeled with:

> area of body/soil
>
> date and time of collection
>
> name of collector
>
> stage *e.g.* larvae, so that if the specimens are pupae when I receive them. I will know that they developed into the next stage during transit.

Handling - most specimens are fairly fragile and are probably best picked up with gloved fingers which are often gentler than forceps if you are not used to them. Very tiny or delicate specimens can be picked up using an artist's brush dipped in water or alcohol depending on what you are about to do with them. Make sure all the vials are very well sealed!

Packing - The insects should be taken to the entomologist as soon as possible. They should be couriered or hand delivered to maintain continuity. They should be packaged in a cardboard box as this has lots of air. Each vial can be taped so that it remains upright. The whole box must remain upright.

I need to know many other factors about the death site :

Habitat -

- o   general - is it woods, a beach, a house, a roadside?
- o   vegetation - trees, grass, bush, shrubs?
- o   soil type - rocky, sandy, muddy?
- o   weather - at time of collection, sunny, cloudy?
- o   temperature and possibly humidity at collection time
- o   elevation and map coordinates of the death site
- o   is the site in shade or direct sunlight?
- o   mention anything unusual, such as whether it's possible that the body may have been submerged at any time.

Remains - I need to know:-

- o   presence, extent and type of clothing
- o   is the body buried or covered? if so, how deep and with what (soil, leaves, cloth)
- o   what is the cause of death, if known? in particular, is there blood at the scene?
- o   or other body fluids?
- o   are there any wounds? if so, what kind?
- o   are drugs likely to be involved? this may affect the decomposition rates
- o   what position is the body in?
- o   what direction is the body facing?
- o   what is the state of decomposition?
- o   Is a maggot mass present? how many? this will affect the temperature on the body
- o   What is the temperature of the centre of the maggot mass(s)?

o   Is there any other meat or carrion around that might also attract insects?
o   Is there a possibility that death did not occur at the present site?

If the body is refrigerated at the morgue before the collection (it is much better to collect at the scene, but sometimes that is not possible) then I also need to know the exact time that the body went into the cooler, and the exact time it came out. **Photographs**, or a video of the scene, the body *in situ* and the site after removal of the body are also extremely useful.

When the insects reach the insectary, the immature specimens are measured, and examined, then placed in a jar containing a suitable feeding media. In the case of blowflies, this is usually beef liver, which is placed on top of sawdust. When the insects reach the prepupal stage and leave the food source they will burrow into the sawdust to pupate. The insects are checked daily and when they pupate they are removed and placed in a petri dish with damp filter paper. The date of pupation and the date of emergence is noted for each specimen. When the adults emerge, they are killed and pinned, then placed in an insect box. Each insect has a detailed label. Any adults collected directly from the corpse are immediately killed and pinned.

The reasons for raising the immature are two-fold. Firstly, larvae are very difficult to identify to species, but adults have many more diagnostic features. Secondly, the dates of pupation and emergence are used to help calculate the age at the time of collection.

Other important information used to determine elapsed time since death, include:

Weather records from the nearest weather station, including temperature and precipitation

The distance between the death site and the weather station

This method of determining elapsed time since death using insect evidence can be demonstrated using an actual case. Human remains were found in mid October. Most of the head region was missing as death was due to gunshot wounds. The upper portion of the body was almost skeletonized, but the lower area, clad in tight clothes, appeared almost fresh. There were several large maggot masses on the corpse which generate their own heat for a while due to the frenzied activity. The temperature of the largest maggot mass was 20°C, even after the body had been refrigerated at 4°C for two hours. All sizes of larvae were collected and three pupae. These were pale in color so had only just pupated. No puparia were found. The mean temperature at the death site was 15°C.

Two species of blowfly emerged, *calliphora vomitoria* and *phormia regina*. Both are common species that are amongst the first to arrive on a corpse. The oldest stage of *calliphora vomitoria* collected was just entering the prepupal stage of the third instar. This was determined from size, no. of spiracular slits (breathing holes) date of pupation and behavior, in that the largest specimens immediately left the beef liver and entered the sawdust, indicating that they had stopped feeding. At the temperature of the death site, 15°C, *calliphora vomitoria* takes a minimum of 9.3 days to reach the beginning of the prepupal stage of the third instar. So these insects were a minimum of 9 days old when collected on 12 October, meaning that they were laid as eggs on or before 4 October. As there was blood at the scene, the insects probably arrived very soon after death. Therefore death must have occurred on or before 4 October.

Using the same techniques for *phormia regina*, the oldest specimens of which were in the pupal stage when collected, it was calculated that *phormia regina* was oviposited no later than 3 October. Therefore, using the two insects together, it can be shown that death occurred on or before 3 October. Other police evidence later showed that death had actually occurred on 3 October.

Post Mortem Interval by Accumulated Degree Hours/Days

Development is regarded as a combination of temperature above the minimum developmental threshold multiplied by time.

- Thus, 5° above the minimum developmental threshold for 2 days = 10 degree days

- And, 1° for 10 days = 10 degree days

- **Degree-day process involves estimating the area under a daily temperature curve that is above the minimum development threshold (generally 6° - 10°C).**

Figure 9.2 Degree-days, a measure of physiological time, indicated as the area under a daily temperature curve and above a minimum developmental temperature for a species.

- With the size of the maggot & which instar it represents known,
- Compare with data from laboratory rearings.
- Then adjust the time period to fit circumstances of the site where the body was found.
- ADH (Accumulated Degree Hours) = time

* temperature

Sample Case

A body is discovered at 8:00a, October 15 and insect specimens are collected & preserved at 9:00am the same day. Most mature maggot of *phormia regina* are molting from first into second instar.

- ADH required = ADH eggs + ADH 1st instar

For laboratory rearings @ 26.7°C

  - time required = 34 hours or 907.8 ADH

- Work backwards from time when maggots were collected, 9:00a, Oct. 15th

  - 9 hours of development that day with mean temp of 20°C; 9*20 = 180 ADH
  - previous day mean temp = 21°C; 21*24 = 504 ADH
  - total time for two days is 180 + 504 = 684 ADH
  - 907.8–684 = 223.8 ADH still required
  - if mean temp on Oct. 13th was 20°C
  - then 223.8 / 20 = 11.2 hours
  - or insect activity initiated between noon and 1pm on October 13th.

Post Mortem Interval by Days

Accumlated Degree Days (ADD)

- Accumulation of heat units above a developmental threshold temperature for an insect larvae.
- **Developmental threshold** (or thermal constant)
- is the minimum temperature necessary for the insect to feed and develop
- A developmental threshold must be known for the insect in question
- Degree Day = Average temperature - Thermal Constant (TC) of insect
- Estimate average temperature (most widely used method)=

   Max temp + Min Temp

DD = 2 - TC

Example of a Green Bottle Blow fly

TC = 7 degrees C
If max. temp = 36 and min = 30 then = 36 + 30 = 66 / 2 = 33 average temp
33 - 7 = 26 degree day

In practice, ADD is worked backwards.

Larvae are collected at the scene and reared to adulthood if it took 6 days in the lab to reach the adult stage at 30 degree days per day and the life cycle developmental data says it takes 300 degree days to reach adulthood

Then 300 - 180 = 120 degree days old when collected.

Then looking at the outside temperature records you can determine how many degree days were possible and pinpoint time of oviposition.

Many common Minimum Developmental Thresholds have been accumulated in (CRC book Forensic Entomology edited by J.H. Byrd and J.L. Castner)

Of course there are influences on the development and succession which are described below.

- Geography
- Seasonal (defined later in this chapter)
- Exposure to sun (even in partial shade)
- Urban vs. rural (City vs. Country)
- Bodies found inside buildings
- Buried bodies
- Bodies in water

- Bodies in vehicles
- Bodies in enclosed spaces
- Hanged bodies
- Burnt remains
- Wrapped remains
- Drugs and other toxins
- Food Type

OTHER USES FOR INSECTS IN FORENSIC SCIENCE

The body may have been moved after death, from the scene of the killing to a hiding place. Some of the insects on the body may be native to the first habitat and not the second. This will show that not only was the body moved, but it will also give an indication of the type of area where the murder actually took place.

The body may have been disturbed after death, by the killer returning to the scene of the crime. This may disturb the insects cycle, and the entomologist may be able to determine not only the date of death, but also the date of the return of the killer.

The presence and position of wounds, decomposition may obscure wounds. Insects colonize remains in a specific pattern, usually laying eggs first in the facial orifices, unless there are wounds, in which case they will colonize these first, then proceed down the body. If the maggot activity is centered away from the natural orifices, then it is likely that this is the site of a wound. For example, maggot activity on the palm of the hands indicates the probable presence of defense wounds.

The presence of drugs can be determined using insect evidence. There is often not enough flesh left to determine drug presence, but maggots bioaccumulate so they can be analyzed to determine type of drug present.

Insects can be used to place a suspect at the scene of a crime. For instance, an insect inside a cocklebur was used to connect a rapist to the rape site. Civil cases also sometimes use insect evidence.

Child or senior abuse/neglect: Some insects will colonize around wounds or unclean areas on a living person. This is called cutaneous myiasis. If the victim is still alive, but maggot infested, a forensic entomologist will be able to tell when the wound or abuse occurred. If there's a case of a neglected child, the onset of maggot infestation will give a minimum time interval since the child last had a diaper change. Such cases occur particularly in young children and seniors. Although forensic entomology can be very effective in determining elapsed time since death, it has its limitations:

The **temperature** of the death site is obviously a very important factor, but few criminals are thoughtful enough to kill their victim right underneath a weather station! In most cases, the weather records come from several miles away. We are trying to overcome this by setting up a miniature weather station at the death site after discovery, to compare these data with that from the weather station, in order to determine the difference between the two sites, if any. Also the microclimate of the corpse itself will be slightly different from the surrounding area, especially if a maggot mass is present. Therefore, it is extremely important to know whether masses are present.

Forensic entomology is **seasonal**, that is, it is only commonly used in spring, summer, and fall when insects are abundant. It is of less use in winter, unless it's very mild, as there are no or very few insects present. This can be a limitation, but can also be an advantage as I can sometimes show that a victim found in spring was killed the previous fall if insect evidence is present.

The results are not immediate, as it takes time to rear the insects. DNA evidence is now being developed to speed up identification of immature specimens.

The body may have been disposed of in a way that **excludes** insects *e.g.*

Freezing - if the body was frozen for a period of time before being placed outside on, for example, 8 May, the insects would only invade then, giving the misleading impression that death had occurred on 8 May. However, other forensic experts would be able to determine whether or not the body has been frozen, and insect evidence will still determine time of exposure.

Burial - if the body is buried deeply, then most insects will be excluded. However, most criminal burials are not very deep, as the aim is merely to conceal the body, and most insects will dig down to the body, particularly if there is blood soaked in the soil. Therefore, insect evidence can still be used.

Wrapped - if the body is wrapped or packaged in some way the insects may be excluded, but the wrapping must be completely secure. A body part was found sealed in a garbage bag which had been tied securely at the top, but the remains were maggot-infested, and showed severe insect damage. The adult females had probably laid their eggs at the knot, and the minute first instar larvae had crawled in.

More **research** is needed. Insect succession varies from geographic region to region and the species and time of colonization must be developed for all areas using this type of evidence. Research has been conducted in British Columbia in a variety of habitats, seasons and geographic areas to develop a database for this Province. It is intended that this will be extended across Canada.

Drugs/Insecticides - the presence of this may affect the development of the insects and, although there may not be a direct correlation as it applies to the atmosphere, a better definition of toxicology follows in this chapter.

In conclusion, **INSECTS ARE EVIDENCE!** Forensic entomology is a very useful method of determining elapsed time since death after 72 h. It is accurate to a day or less, or a range of days, and may be the only method available to determine elapsed time since death. It is vital that the insects are collected properly and its accuracy depends on this and on suitable conditions for insects.

Chapter 6

FORENSIC TOXICOLOGY

Forensic toxicology is essentially a specialty area of analytical chemistry. *Toxicology* is the science of adverse effects of chemicals on living organisms. In general, a toxicologist detects and identifies foreign chemicals in the body, with a particular emphasis upon toxic or hazardous substances. A *descriptive toxicologist* performs toxicity tests to evaluate the risk that exposure poses to humans. A *mechanistic toxicologist* attempts to determine how substances exert deleterious effects on living organisms. A *regulatory toxicologist* judges whether or not a substance has low enough risk to justify making it available to the public.

A *toxin* is any material exerting a life threatening effect upon a living organism. Poisons are a subgroup of toxins. Toxic materials exist in many forms (gaseous, liquid, solid, animal, mineral, and vegetable), and may be ingested, inhaled, or absorbed through the skin. *Poisons* generally enter the body in a single massive dose, or accumulate to a massive dose over time. Toxins work in minute quantities or low levels, requiring sensitive analytical instruments for detection. Some toxins have medicinal value, but many produce irreparable damage. Some toxins have antidotes and others do not. Poisons can be combated by prompt treatment, and most organ damage (except for serious CNS injury) may be repairable. Whereas poisons are somewhat easily identifiable by their symptoms, many toxins tend to disguise or mask themselves.

The toxic effects of substances are *not* side effects. Side effects are defined as non-deleterious, such as dry mouth, for example. Toxic effects are the undesirable results of a direct effect. They occur in a number of ways, most often produced by a dangerous metabolite of the drug which is activated by an enzyme, light, or oxygen reaction in a process known as *biotransformation*. Toxic reactions often depend on how metabolites are processed by an individual's body, how proteins build up and bind at effector sites in the body. Some metabolites destroy liver cells, others brain tissue, and still others operate at the DNA level. Toxic reactions are classified as one of three (3) reactions:

- pharmacological—injury to the central nervous system (CNS)
- pathological—injury to the liver
- genotoxic—creation of benign or malignant neoplasms or tumors

If the concentration of toxin doesn't reach a *critical level*, the effects will usually be reversible. Pharmacological reactions, for example, are of this type. In order to sustain permanent brain damage, dosages must be above a standard critical level. Pathological reactions can be repaired if discovered early enough, but most liver damage occurs over a period of few months to a decade. Genotoxic or carcinogenic effects may take 20–40 years before tumors develop. Most of the time, toxic metabolites are activated by enzymatic transformation, but a few are activated by light. This means that exposure of the skin to sunlight produces a photoallergic reaction or *phototoxic reaction* within 24 hours. It's important to understand that *the target organ of toxicity is not the site where toxin accumulates.*

Lead poisoning, for example, results in an accumulation of lead in bone marrow, but the toxic effect is the creation of lesions on skin and soft tissue. Carcinogenesis is even more complicated, involving the creation of promotor electrophiles which serve to activate or potentiate the growth of latent tumors given some biological trigger or subsequent environmental attack. Different people, of course, have *chemical allergies* (as well as food allergies), depending upon the serology of their allergen-antigen history. In such people, toxic reactions take different forms. Other people have what are called *idosyncratic reactions*, which means they have certain unique genetic triggers. Furthermore, people exposed to multiple toxins can have *synergistic reactions*, which means that two or more toxins interact at the metabolic level to be greater or less than the effects of the individual toxins.

A forensic toxicologist is normally presented with preserved samples of body fluids, stomach contents, and organ parts. They will have access to the coroner's report which should contain information on various signs and symptoms as well as other postmortem data. The toxicologist needs a through knowledge of how the body alters or metabolizes drugs because few substances leave the body in the same state as they entered. The substances they work with are often *derivatives*, which is a term meaning a chemical compound which is prepared from a pure compound in order to be more easily detected by the analytical techniques used. They also divide specimens up into acidic and basic *fractions* for drug extraction from tissue or fluid. Almost all drugs are either acids or bases (on a pH scale from 0 to 14 with closer to 0 being acids and closer to 14 being bases). Acid drugs are easily extracted with a pH solution of less than 7; base drugs are easily extracted with a pH solution of greater than 7. As an example, most of the barbiturate drugs are acid-soluble; most of the amphetamine drugs are base-soluble.

After preliminary acid-base procedures are carried out, and the tissue or fluid sample is now a drug sample, examination continues in two steps: (1) screening tests, and (2) confirmation testing. Screening tests allow the processing of many specimens for a wide range of toxins in a short time. Any positive indications from the screening tests must be verified with a confirmation test. The following are some standard laboratory tests for toxin detection:

SCREENING TESTS

- Physical tests—boiling point, melting point, density, and refractive index
- Crystal tests—treatment with a chemical reagent to produce crystals
- Chemical spot tests—treatment with a chemical reagent to produce color changes
- Chromatography (thin-layer or gas)—used to separate components of a mixture

CONFIRMATION TESTS

- Mass spectrometry—this is a combination of gas chromatography/mass spectrometry which is generally accepted as the confirmation test of choice. Each toxin has a known mass spectra, or "fingerprint", which is infallible proof of its presence at the chemical level

Ordinarily, the toxicologist is not required to render an opinion of whether the toxin levels in the body were enough to cause death. A few toxicologists may do so, but they must have had special training in physiology, and this is usually the province of the forensic pathologist, in any event. Often, the defense will call their medical experts to dispute a cause of death claim. Physicians are the only ones qualified to render opinions on the physiological effects of toxins, and forensic law allows them to provide their testimony in the form of *hypotheticals*, even though they do not have personal knowledge of the case. Low-level toxin cases usually become a real battle of the experts.

Drug overdoses and alcoholic poisonings will provide most of the work for toxicologists; hence a couple of allied subfields may be drawn upon: (1) a field inhabited by what are called Drug Recognition Experts (DRE); and (2) alcohol intoxication measurement (a subject talked about in a previous lecture). Both are sought-after areas of police training. Another related subfield involves carbon monoxide poisoning, which may involve an automobile engineer or fire safety specialist.

The Drug Recognition Expertise evolved out of experiments in California with the LAPD during the 1970s in which police officers were trained to identify and recognize certain types of drugs based upon the impairments and physiological symptoms. The examination that such specially trained police officers conduct goes beyond normal Nystagmus testing and more closely resembles the taking of vital signs by a nurse or paramedic, combined with structured interviewing and observation.

Benjamin, D. (1993). "Forensic Pharmacology" in R. Saferstein (ed.) *Forensic Science Handbook*. NJ: Prentice-Hall.

Klaasen, C. (1996). "Principles of Toxicology and Treatment of Poisoning" in J. Hardman et al., *Goodman and Gilman's The Pharmacological Basis of Therapeutics*. NY: McGraw-Hill.

Levine, A. (1993). "Forensic Toxicology" *Journal of Analytical Chemistry* 65: 272–76.

Lowry, W. & J. Garriott. (1979). *Forensic Toxicology: Controlled Substances and Dangerous Drugs*. NY: Plenum.

Moenssens, A.A.; Inbau, F.E.; Starrs, J.E. (1986). Scientific Evidence in Civil and Criminal Cases. NY: The Foundation Press.

Saferstein, R. (1998). *Criminalistics: An Introduction to Forensic Science*. NJ: Prentice-Hall.

Chapter 7

NATIONAL GUIDELINES FOR DEATH INVESTIGATION

In June and July 1997, NMRP met for two 1 1/2-day working sessions in St. Louis and Chicago to review the draft guidelines developed by the executive board and offer recommendations and changes based on jurisdictional variances and organizational responsibilities. Those sessions resulted in the final draft of the 29 guidelines for conducting death investigations. The 29 guidelines are presented in the next main section.

Guideline Status

Currently, NMRP members are presenting the guidelines to their respective organizations' leadership (or appropriate internal committees) for review. This researcher is collecting anecdotal comments for future modification of the existing guidelines during the validation procedures.

Training Guidelines

The purpose of the second part of the national death investigator guidelines research was to identify training criteria for each of the 29 guidelines. This research is now completed. (A death investigation curriculum guide and a CD-ROM for use in the field or classroom are in development.) For each of the guidelines presented in this report, "minimum levels of performance" will be developed and verified by the members of the various TWGs. These "training guidelines" will provide both individuals and educational organizations the material needed to establish and maintain valid exit outcomes for each investigative trainee.

Guideline Validation

In this initial research, 29 investigative tasks were identified. Each task was developed into a guideline for investigators to follow while conducting a death investigation. Although each TWG believed in the validity of each guideline, no attempt was made to validate actual significance (e.g., if guideline C1 is trained and implemented, a [%] decrease in poor scene photographs should occur). The researcher is currently developing a national validation strategy for the implementation and validation of each guideline.

This handbook is intended as a guide to recommended practices for the investigation of death scenes. Jurisdictional, logistical, or legal conditions may preclude the use of particular procedures contained herein.

Section A: Investigative Tools and Equipment

1. Gloves (universal precautions).
2. Writing implements (pens, pencils, markers).
3. Body bags.
4. Communication equipment (cell phone, pager, radio).
5. Flashlight.
6. Body ID tags.
7. Camera—35mm (with extra batteries, film, etc.).
8. Investigative notebook (for scene notes, etc.).
9. Measurement instruments (tape measure, ruler, rolling measuring tape, etc.).
10. Official identification (for yourself).
11. Watch.
12. Paper bags (for hands, feet, etc.).
13. Specimen containers (for evidence items and toxicology specimens).
14. Disinfectant (universal Pprecautions).
15. Departmental scene forms.

16. Camera—Polaroid (with extra film).
17. Blood collection tubes (syringes and needles).
18. Inventory lists (clothes, drugs, etc.).
19. Paper envelopes.
20. Clean white linen sheet (stored in plastic bag).
21. Evidence tape.
22. Business cards/office cards w/phone numbers.
23. Foul-weather gear (raincoat, umbrella, etc.).
24. Medical equipment kit (scissors, forceps, tweezers, exposure suit, scalpel handle, blades, disposable syringe, large gauge needles, cotton-tipped swabs, etc.).
25. Phone listing (important phone numbers).
26. Tape or rubber bands.
27. Disposable (paper) jumpsuits, hair covers, face shield, etc.
28. Evidence seal (use with body bags/locks).
29. Pocketknife.
30. Shoe-covers.
31. Trace evidence kit (tape, etc.).
32. Waterless hand wash.
33. Thermometer.
34. Crime scene tape.
35. First aid kit.
36. Latent print kit.
37. Local maps.
38. Plastic trash bags.
39. Gunshot residue analysis kits (SEM/EDS).
40. Photo placards (signage to ID case in photo).
41. Boots (for wet conditions, construction sites, etc.).
42. Hand lens (magnifying glass).
43. Portable electric area lighting.
44. Barrier sheeting (to shield body/area from public view).
45. Purification mask (disposable).
46. Reflective vest.
47. Tape recorder.

48. Basic handtools (boltcutter, screwdrivers, hammer, shovel, trowel, paintbrushes, etc.).
49. Body bag locks (to secure body inside bag).
50. Camera—Video (with extra battery).
51. Personal comfort supplies (insect spray, sun screen, hat, etc.).
52. Presumptive blood test kit.

Section B: Arriving at the Scene

1. Introduce and Identify Self and Role

   Principle: Introductions at the scene allow the investigator to establish formal contact with other official agency representatives. The investigator must identify the first responder to ascertain if any artifacts or contamination may have been introduced to the death scene. The investigator must work with all key people to ensure scene safety prior to his/her entrance into the scene.

   Authorization: Medical Examiner/Coroner Official Office Policy Manual; State or Federal Statutory Authority.

   Policy: The investigator shall take the initiative to introduce himself or herself, identify essential personnel, establish rapport, and determine scene safety.

   Procedure: Upon arrival at the scene, and prior to entering the scene, the investigator should:

   A. Identify the lead investigator at the scene and present identification.

B. Identify other essential officials at the scene (e.g., law enforcement, fire, EMS, social/child protective services, etc.) and explain the investigator's role in the investigation.
C. Identify and document the identity of the first essential official(s) to the scene (first "professional" arrival at the scene for investigative followup) to ascertain if any artifacts or contamination may have been introduced to the death scene.
D. Determine the scene safety (prior to entry).

Summary: Introductions at the scene help to establish a collaborative investigative effort. It is essential to carry identification in the event of questioned authority. It is essential to establish scene safety prior to entry.

2. Exercise Scene Safety

Principle: Determining scene safety for all investigative personnel is essential to the investigative process. The risk of environmental and physical injury must be removed prior to initiating a scene investigation. Risks can include hostile crowds, collapsing structures, traffic, and environmental and chemical threats.

Authorization: Medical Examiner/Coroner Official Office Policy Manual; State or Federal Statutory Authority.

Policy: The investigator shall attempt to establish scene safety prior to entering the scene to prevent injury or loss of life, including contacting appropriate agencies for assistance with other scene safety issues.

Procedure: Upon arrival at the scene, the investigator should:

A. Assess and/or establish physical boundaries.
B. Identify incident command.
C. Secure vehicle and park as safely as possible.
D. Use personal protective safety devices (physical, biochemical safety).
E. Arrange for removal of animals or secure (if present and possible).
F. Obtain clearance/authorization to enter scene from the individual responsible for scene safety (e.g., fire marshal, disaster coordinator).
G. While exercising scene safety, protect the integrity of the scene and evidence to the extent possible from contamination or loss by people, animals, and elements.

Note: Due to potential scene hazards (e.g., crowd control, collapsing structures, poisonous gases, traffic), the body may have to be removed before scene investigation can be continued.

Summary: Environmental and physical threats to the investigator must be removed in order to conduct a scene investigation safely. Protective devices must be used by investigative staff to prevent injury. The investigator must endeavor to protect the evidence against contamination or loss.

3. Confirm or Pronounce Death

   Principle: Appropriate personnel must make a determination of death prior to the initiation of the death investigation. The confirmation or pronouncement of death determines jurisdictional responsibilities.

   Authorization: Medical Examiner/Coroner Official Office Policy Manual; State or Federal Statutory Authority.

   Policy: The investigator shall ensure that appropriate personnel have viewed the body and that death has been confirmed.

   Procedure: Upon arrival at the scene, the investigator should:

   A. Locate and view the body.
   B. Check for pulse, respiration, and reflexes, as appropriate.
   C. Identify and document the individual who made the official determination of death, including the date and time of determination.
   D. Ensure death is pronounced, as required.

   Summary: Once death has been determined, rescue/resuscitative efforts cease and medico-legal jurisdiction can be established. It is vital that this occur prior to the medical examiner/coroner's assuming any responsibilities.

4. Participate in Scene Briefing (With Attending Agency Representatives)

Principle: Scene investigators must recognize the varying jurisdictional and statutory responsibilities that apply to individual agency representatives (e.g., law enforcement, fire, EMT, judicial/legal). Determining each agency's investigative responsibility at the scene is essential in planning the scope and depth of each scene investigation and the release of information to the public.

Authorization: Medical Examiner/Coroner Official Office Policy Manual; State or Federal Statutory Authority.

Policy: The investigator shall identify specific responsibilities, share appropriate preliminary information, and establish investigative goals of each agency present at the scene.

Procedure: When participating in scene briefing, the investigator should:

   A. Locate the staging area (entry point to scene, command post, etc.).
   B. Document the scene location (address, mile marker, building name) consistent with other agencies.
   C. Determine nature and scope of investigation by obtaining preliminary investigative details (e.g., suspicious versus nonsuspicious death).
   D. Ensure that initial accounts of incident are obtained from the first witness(es).

Summary: Scene briefing allows for initial and factual information exchange. This includes scene location, time factors, initial witness information, agency responsibilities, and investigative strategy.

5. Conduct Scene "Walk Through"

   Principle: Conducting a scene "walk through" provides the investigator with an overview of the entire scene. The "walk through" provides the investigator with the first opportunity to locate and view the body, identify valuable and/or fragile evidence, and determine initial investigative procedures providing for a systematic examination and documentation of the scene and body.

   Authorization: Medical Examiner/Coroner Official Office Policy Manual; State or Federal Statutory Authority.

   Policy: The investigator shall conduct a scene "walk through" to establish pertinent scene parameters.

   Procedure: Upon arrival at the scene, the investigator should:

   A. Reassess scene boundaries and adjust as appropriate.
   B. Establish a path of entry and exit.
   C. Identify visible physical and fragile evidence.
   D. Document and photograph fragile evidence immediately and collect if appropriate.
   E. Locate and view the decedent.

Summary: The initial scene "walk through" is essential to minimize scene disturbance and to prevent the loss and/or contamination of physical and fragile evidence.

6. Establish Chain of Custody

Principle: Ensuring the integrity of the evidence by establishing and maintaining a chain of custody is vital to an investigation. This will safeguard against subsequent allegations of tampering, theft, planting, and contamination of evidence.

Authorization: Medical Examiner/Coroner Official Office Policy Manual; State or Federal Statutory Authority.

Policy: Prior to the removal of any evidence, the custodian(s) of evidence shall be designated and shall generate and maintain a chain of custody for all evidence collected.

Procedure: Throughout the investigation, those responsible for preserving the chain of custody should:

    A. Document location of the scene and time of arrival of the death investigator at the scene.
    B. Determine custodian(s) of evidence, determine which agency(ies) is/are responsible for collection of specific types of evidence, and determine evidence collection priority for fragile/fleeting evidence.

C. Identify, secure, and preserve evidence with proper containers, labels, and preservatives.
   This is the time when a four sample reading of either rectal or liver temperatures would be needed. (4–10 minute readings logging body temp, air temp and ground temp). This information would be beneficial when entering the data into the new ETOD computer program that is currently in development at Cyclogenesis.
D. Document the collection of evidence by recording its location at the scene, time of collection, and time and location of disposition.
E. Develop personnel lists, witness lists, and documentation of times of arrival and departure of personnel.

Summary:

It is essential to maintain a proper chain of custody for evidence. Through proper documentation, collection, and preservation, the integrity of the evidence can be assured. A properly maintained chain of custody and prompt transfer will reduce the likelihood of a challenge to the integrity of the evidence.

7. Follow Laws (Related to the Collection of Evidence)

Principle: The investigator must follow local, State, and Federal laws for the collection of evidence to ensure its admissibility. The investigator must work with law enforcement and the legal authorities to determine laws regarding collection of evidence.

Authorization: Medical Examiner/Coroner Official Office Policy Manual; State or Federal Statutory Authority.

Policy: The investigator working with other agencies must identify and work under appropriate legal authority. Modification of informal procedures may be necessary but laws must always be followed.

Procedure: The investigator, prior to or upon arrival at the death scene, should work with other agencies to:

A. Determine the need for a search warrant (discuss with appropriate agencies).
B. Identify local, State, Federal, and international laws (discuss with appropriate agencies).
C. Identify medical examiner/coroner statutes and/or office standard operating procedures (discuss with appropriate agencies).

Summary: Following laws related to the collection of evidence will ensure a complete and proper investigation in compliance with State and local laws, admissibility in court, and adherence to office policies and protocols.

Section C: Documenting and Evaluating the Scene

1. Photograph Scene

Principle: The photographic documentation of the scene creates a permanent historical record of the scene. Photographs provide detailed corroborating evidence that constructs a system of redundancy should questions arise concerning the report, witness statements, or position of evidence at the scene.

Authorization: Medical Examiner/Coroner Official Office Policy Manual; State or Federal Statutory Authority.

Policy: The investigator shall obtain detailed photographic documentation of the scene that provides both instant and permanent high-quality (e.g., 35 mm) images.

Procedure: Upon arrival at the scene, and prior to moving the body or evidence, the investigator should:

- A. Remove all nonessential personnel from the scene.
- B. Obtain an overall (wide-angle) view of the scene to spatially locate the specific scene to the surrounding area.
- C. Photograph specific areas of the scene to provide more detailed views of specific areas within the larger scene.
- D. Photograph the scene from different angles to provide various perspectives that may uncover additional evidence.
- E. Obtain some photographs with scales to document specific evidence.
- F. Obtain photographs even if the body or other evidence has been moved.

Note: If evidence has been moved prior to photography, it should be noted in the report, but the body or other evidence should not be reintroduced into the scene in order to take photographs.

Summary: Photography allows for the best permanent documentation of the death scene. It is essential that accurate scene photographs are available for other investigators, agencies, and authorities to recreate the scene. Photographs are a permanent record of the terminal event and retain evidentiary value and authenticity. It is essential that the investigator obtain accurate photographs before releasing the scene.

2. Develop Descriptive Documentation of the Scene

Principle: Written documentation of the scene(s) provides a permanent record that may be used to correlate with and enhance photographic documentation, refresh recollections, and record observations.

Authorization: Medical Examiner/Coroner Official Office Policy Manual; State or Federal Statutory Authority.

Policy: Investigators shall provide written scene documentation.

Procedure: After photographic documentation of the scene and prior to removal of the body or other evidence, the investigator should:

A. Diagram/describe in writing items of evidence and their relationship to the body with necessary measurements.
B. Describe and document, with necessary measurements, blood and body fluid evidence including volume, patterns, spatters, and other characteristics.
C. Describe scene environments including odors, lights, temperatures, and other fragile evidence.

Note: If evidence has been moved prior to written documentation, it should be noted in the report.

Summary: Written scene documentation is essential to correlate with photographic evidence and to recreate the scene for police, forensic(s), and judicial and civil agencies with a legitimate interest.

3. Establish Probable Location of Injury or Illness

Principle: The location where the decedent is found may not be the actual location where the injury/illness that contributed to the death occurred. It is imperative that the investigator attempt to determine the locations of any and all injury(ies)/illness(es) that may have contributed to the death.

Physical evidence at any and all locations may be pertinent in establishing the cause, manner, and circumstances of death.

Authorization: Medical Examiner/Coroner Official Office Policy Manual; State or Federal Statutory Authority.

Policy: The investigator shall obtain detailed information regarding any and all probable locations associated with the individual's death.

Procedure: The investigator should:

A. Document location where death was confirmed.
B. Determine location from which decedent was transported and how body was transported to scene.
C. Identify and record discrepancies in rigor mortis, livor mortis, and body temperature.
D. Check body, clothing, and scene for consistency/inconsistency of trace evidence and indicate location where artifacts are found.
E. Check for drag marks (on body and ground).
F. Establish post-injury activity.
G. Obtain dispatch (e.g., police, ambulance) record(s).
H. Interview family members and associates as needed.

Summary: Due to post-injury survival, advances in emergency medical services, multiple modes of transportation, the availability of specialized care, or criminal activity, a body may be moved from the actual location of illness/injury to a remote site. It is imperative that the investigator attempt to determine any and all locations where the decedent has previously been and the mode of transport from these sites.

4. Collect, Inventory, and Safeguard Property and Evidence

   Principle: The decedent's valuables/property must be safeguarded to ensure proper processing and eventual return to next of kin. Evidence on or near the body must be safeguarded to ensure its availability for further evaluation.

   Authorization: Medical Examiner/Coroner Official Office Policy Manual; State or Federal Statutory Authority.

   Policy: The investigator shall ensure that all property and evidence is collected, inventoried, safeguarded, and released as required by law.

   Procedure: After personal property and evidence have been identified at the scene, the investigator (with a witness) should:

   A. Inventory, collect, and safeguard illicit drugs and paraphernalia at scene and/or office.
   B. Inventory, collect, and safeguard prescription medication at scene and/or office.
   C. Inventory, collect, and safeguard over-the-counter medications at scene and/or office.
   D. Inventory, collect, and safeguard money at scene and at office.

E. Inventory, collect, and safeguard personal valuables /property at scene and at office.

Summary: Personal property and evidence are important items at a death investigation. Evidence must be safeguarded to ensure its availability if needed for future evaluation and litigation. Personal property must be safeguarded to ensure its eventual distribution to appropriate agencies or individuals and to reduce the likelihood that the investigator will be accused of stealing property.

5. Interview Witness(es) at the Scene

Principle: The documented comments of witnesses at the scene allow the investigator to obtain primary source data regarding discovery of body, witness corroboration, and terminal history. The documented interview provides essential information for the investigative process.

Authorization: Medical Examiner/Coroner Official Office Policy Manual; State or Federal Statutory Authority.

Policy: The investigator's report shall include the source of information, including specific statements and information provided by the witness.

Procedure: Upon arriving at the scene, the investigator should:

A. Collect all available identifying data on witnesses (e.g., full name, address,

    DOB, work and home telephone numbers, etc.).
 B. Establish witness' relationship / association to the deceased.
 C. Establish the basis of witness' knowledge (how does witness have knowledge of the death?).
 D. Obtain information from each witness.
 E. Note discrepancies from the scene briefing (challenge, explain, verify statements).
 F. Tape statements where such equipment is available and retain them.

Summary: The final report must document witness' identity and must include a summary of witness' statements, corroboration with other witnesses, and the circumstances of discovery of the death. This documentation must exist as a permanent record to establish a chain of events.

Section D: Documenting and Evaluating the Body

1. Photograph the Body

    Principle: The photographic documentation of the body at the scene creates a permanent record that preserves essential details of the body position, appearance, identity, and final movements. Photographs allow sharing of information with other agencies investigating the death.

    Authorization: Medical Examiner/Coroner Official Office Policy Manual; State or Federal Statutory Authority.

Policy: The investigator shall obtain detailed photographic documentation of the body that provides both instant and permanent high-quality (e.g., 35 mm) images.

Procedure: Upon arrival at the scene, and prior to moving the body or evidence, the investigator should:

    A. Photograph the body and immediate scene (including the decedent as initially found).
    B. Photograph the decedent's face.
    C. Take additional photographs after removal of objects/items that interfere with photographic documentation of the decedent (e.g., body removed from car).
    D. Photograph the decedent with and without measurements (as appropriate).
    E. Photograph the surface beneath the body (after the body has been removed, as appropriate).

Note: Never clean face, do not change condition. Take multiple shots if possible.

Summary: The photographic documentation of the body at the scene provides for documentation of the body position, identity, and appearance. The details of the body at the scene provide investigators with pertinent information of the terminal events.

2. Conduct External Body Examination (Superficial)

   Principle: Conducting the external body examination provides the investigator with objective data regarding the single most important piece of evidence at the scene, the body. This documentation provides detailed information regarding the decedent's physical attributes, his/her relationship to the scene, and possible cause, manner, and circumstances of death.

   Authorization: Medical Examiner/Coroner Official Office Policy Manual; State or Federal Statutory Authority.

   Policy: The investigator shall obtain detailed photographs and written documentation of the decedent at the scene.

   Procedure: After arrival at the scene and prior to moving the decedent, the investigator should, without removing decedent's clothing:

   A. Photograph the scene, including the decedent as initially found and the surface beneath the body after the body has been removed.

   Note: If necessary, take additional photographs after removal of objects/items that interfere with photographic documentation of the decedent.

B. Photograph the decedent with and without measurements (as appropriate), including a photograph of the decedent's face.
C. Document the decedent's position with and without measurements (as appropriate).
D. Document the decedent's physical characteristics.
E. Document the presence or absence of clothing and personal effects.
F. Document the presence or absence of any items/objects that may be relevant.
G. Document the presence or absence of marks, scars, and tattoos.
H. Document the presence or absence of injury/trauma, petechiae, etc.
I. Document the presence of treatment or resuscitative efforts.
J. Based on the findings, determine the need for further evaluation/assistance of forensic specialists (e.g., pathologists, odontologists).

Summary: Thorough evaluation and documentation (photographic and written) of the deceased at the scene is essential to determine the depth and direction the investigation will take.

3. Preserve Evidence (on Body)

Principle: The photographic and written documentation of evidence on the body allows the investigator to obtain a permanent historical record of that evidence. To maintain chain of custody, evidence must be collected, preserved, and transported properly. In addition to all of the physical evidence visible on the body, blood and other body fluids present must be photographed and documented prior to collection and transport. Fragile evidence (that which can be easily contaminated, lost, or altered) must also be collected and/or preserved to maintain chain of custody and to assist in determination of cause, manner, and circumstances of death.

Authorization: Medical Examiner/Coroner Official Office Policy Manual; State or Federal Statutory Authority.

Policy: With photographic and written documentation, the investigator will provide a permanent record of evidence that is on the body.

Procedure: Once evidence on the body is recognized, the investigator should:

A. Photograph the evidence.
B. Document blood/body fluid on the body (froth/purge, substances from orifices), location, and pattern before transporting.
C. Place decedent's hands and/or feet in unused paper bags (as determined by the scene).
D. Collect trace evidence before transporting the body (e.g., blood, hair, fibers, etc.).
E. Arrange for the collection and transport of evidence at the scene (when necessary).

F. Ensure the proper collection of blood and body fluids for subsequent analysis (if body will be released from scene to an outside agency without an autopsy).

Summary: It is essential that evidence be collected, preserved, transported, and documented in an orderly and proper fashion to ensure the chain of custody and admissibility in a legal action. The preservation and documentation of the evidence on the body must be initiated by the investigator at the scene to prevent alterations or contamination.

4. Establish Decedent Identification

Principle: The establishment or confirmation of the decedent's identity is paramount to the death investigation. Proper identification allows notification of next of kin, settlement of estates, resolution of criminal and civil litigation, and the proper completion of the death certificate.

Authorization: Medical Examiner/Coroner Official Office Policy Manual; State or Federal Statutory Authority.

Policy: The investigator shall engage in a diligent effort to establish/confirm the decedent's identity.

Procedure: To establish identity, the investigator should document use of the following methods:

A. Direct visual or photographic identification of the decedent if visually recognizable.
B. Scientific methods such as fingerprints, dental, radiographic, and DNA comparisons.
C. Circumstantial methods such as (but not restricted to) personal effects, circumstances, physical characteristics, tattoos, and anthropologic data.

Summary: There are several methods available that can be used to properly identify deceased persons. This is essential for investigative, judicial, family, and vital records issues.

5. Document Post Mortem Changes

Principle: The documenting of post mortem changes to the body assists the investigator in explaining body appearance in the interval following death. Inconsistencies between post mortem changes and body location may indicate movement of body and validate or invalidate witness statements. In addition, post mortem changes to the body, when correlated with circumstantial information, can assist the investigators in estimating the approximate time of death.

Authorization: Medical Examiner/Coroner Official Office Policy Manual; State or Federal Statutory Authority.

Policy: The investigator shall document all post mortem changes relative to the decedent and the environment.

Procedure: Upon arrival at the scene and prior to moving the body, the investigator should note the presence of each of the following in his/her report:

A. Livor (color, location, blanchability, Tardieu spots) consistent/inconsistent with position of the body.
B. Rigor (stage/intensity, location on the body, broken, inconsistent with the scene).
C. Degree of decomposition (putrefaction, adipocere, mummification, skeletonization, as appropriate).
D. Insect and animal activity.
E. Scene temperature (document method used and time estimated).
F. Description of body temperature (e.g., warm, cold, frozen) or measurement of body temperature (document method used and time of measurement).

Summary: Documentation of post mortem changes in every report is essential to determine an accurate cause and manner of death, provide information as to the time of death, corroborate witness statements, and indicate that the body may have been moved after death.

6. Participate in Scene Debriefing

Principle: The scene debriefing helps investigators from all participating agencies to establish post-scene responsibilities by sharing data regarding particular scene findings. The scene debriefing provides each agency the opportunity for input regarding special requests for assistance, additional information, special examinations, and other requests requiring interagency communication, cooperation, and education.

Authorization: Medical Examiner/Coroner Official Office Policy Manual; State or Federal Statutory Authority.

Policy: The investigator shall participate in or initiate interagency scene debriefing to verify specific post-scene responsibilities.

Procedure: When participating in scene debriefing, the investigator should:

A. Determine post-scene responsibilities (identification, notification, press relations, and evidence transportation).
B. Determine/identify the need for a specialist (e.g., crime laboratory technicians, social services, entomologists, OSHA).
C. Communicate with the pathologist about responding to the scene or to the autopsy schedule (as needed).
D. Share investigative data (as required in furtherance of the investigation).
E. Communicate special requests to appropriate agencies, being mindful of the necessity for confidentiality.

Summary: The scene debriefing is the best opportunity for investigative participants to communicate special requests and confirm all current and additional scene responsibilities. The debriefing allows participants the opportunity to establish clear lines of responsibility for a successful investigation.

7. Determine Notification Procedures (Next of Kin)

   Principle: Every reasonable effort should be made to notify the next of kin as soon as possible. Notification of next of kin initiates closure for the family, disposition of remains, and facilitates the collection of additional information relative to the case.

   Authorization: Medical Examiner/Coroner Official Office Policy Manual; State or Federal Statutory Authority.

   Policy: The investigator shall ensure that next of kin is notified of the death and that all failed and successful attempts at notification are documented.

   Procedure: When determining notification procedures, the investigator should:

   A. Identify next of kin (determine who will perform task).
   B. Locate next of kin (determine who will perform task).
   C. Notify next of kin (assign person(s) to perform task) and record time of notification, or, if delegated to another agency, gain confirmation when notification is made.
   D. Notify concerned agencies of status of the notification.

Summary: The investigator is responsible for ensuring that the next of kin is identified, located, and notified in a timely manner. The time and method of notification should be documented. Failure to locate next of kin and efforts to do so should be a matter of record. This ensures that every reasonable effort has been made to contact the family.

8. Ensure Security of Remains

   Principle: Ensuring security of the body requires the investigator to supervise the labeling, packaging, and removal of the remains. An appropriate identification tag is placed on the body to preclude misidentification upon receipt at the examining agency. This function also includes safeguarding all potential physical evidence and/or property and clothing that remain on the body.

   Authorization: Medical Examiner/Coroner Official Office Policy Manual; State or Federal Statutory Authority.

   Policy: The investigator shall supervise and ensure the proper identification, inventory, and security of evidence/property and its packaging and removal from the scene.

   Procedure: Prior to leaving the scene, the investigator should:

   A. Ensure that the body is protected from further trauma or contamination (if not, document) and unauthorized

removal of therapeutic and resuscitative equipment.
B. Inventory and secure property, clothing, and personal effects that are on the body (remove in a controlled environment with witness present).
C. Identify property and clothing to be retained as evidence (in a controlled environment).
D. Recover blood and/or vitreous samples prior to release of remains.
E. Place identification on the body and body bag.
F. Ensure/supervise the placement of the body into the bag.
G. Ensure/supervise the removal of the body from the scene.
H. Secure transportation.

Summary: Ensuring the security of the remains facilitates proper identification of the remains, maintains a proper chain of custody, and safeguards property and evidence.

Section E: Establishing and Recording Decedent Profile Information

1. Document the Discovery History

Principle: Establishing a decedent profile includes documenting a discovery history and circumstances surrounding the discovery. The basic profile will dictate subsequent levels of investigation, jurisdiction, and authority. The focus (breadth/depth) of further investigation is dependent on this information.

Authorization: Medical Examiner/Coroner Official Office Policy Manual; State or Federal Statutory Authority.

Policy: The investigator shall document the discovery history, available witnesses, and apparent circumstances leading to death.

Procedure: For an investigator to correctly document the discovery history, he/she should:

  A. Establish and record person(s) who discovered the body and when.
  B. Document the circumstances surrounding the discovery (who, what, where, when, how).

Summary: The investigator must produce clear, concise, documented information concerning who discovered the body, what are the circumstances of discovery, where the discovery occurred, when the discovery was made, and how the discovery was made.

2. Determine Terminal Episode History

Principle: Pre-terminal circumstances play a significant role in determining cause and manner of death. Documentation of medical intervention and/or procurement of ante mortem specimens help to establish the decedent's condition prior to death.

Authorization: Medical Examiner/Coroner Official Office Policy Manual; State or Federal Statutory Authority.

Policy: The investigator shall document known circumstances and medical intervention preceding death.

Procedure: In order for the investigator to determine terminal episode history, he/she should:

A. Document when, where, how, and by whom decedent was last known to be alive.
B. Document the incidents prior to the death.
C. Document complaints/symptoms prior to the death.
D. Document and review complete EMS records (including the initial electrocardiogram).
E. Obtain relevant medical records (copies).
F. Obtain relevant ante mortem specimens.

Summary: Obtaining records of pre-terminal circumstances and medical history distinguishes medical treatment from trauma. This history and relevant ante mortem specimens assist the medical examiner/coroner in determining cause and manner of death.

3. Document Decedent Medical History

Principle: The majority of deaths referred to the medical examiner/coroner are natural deaths. Establishing the decedent's medical history helps to focus the investigation. Documenting the decedent's medical signs or symptoms prior to death determines the need for subsequent examinations.

The relationship between disease and injury may play a role in the cause, manner, and circumstances of death.

Authorization: Medical Examiner/Coroner Official Office Policy Manual; State or Federal Statutory Authority.

Policy: The investigator shall obtain the decedent's past medical history.

Procedure: Through interviews and review of the written records, the investigator should:

    A. Document medical history, including medications taken, alcohol and drug use, and family medical history from family members and witnesses.
    B. Document information from treating physicians and/or hospitals to confirm history and treatment.
    C. Document physical characteristics and traits (e.g., left-/right-handedness, missing appendages, tattoos, etc.).

Summary: Obtaining a thorough medical history focuses the investigation, aids in disposition of the case, and helps determine the need for a post mortem examination or other laboratory tests or studies.

4. Document Decedent Mental Health History

Principle: The decedent's mental health history can provide insight into the behavior/state of mind of the individual. That insight may produce clues that will aid in establishing the cause, manner, and circumstances of the death.

Authorization: Medical Examiner/Coroner Official Office Policy Manual; State or Federal Statutory Authority.

Policy: The investigator shall obtain information from sources familiar with the decedent pertaining to the decedent's mental health history.

Procedure: The investigator should:

A. Document the decedent's mental health history, including hospitalizations and medications.
B. Document the history of suicidal ideations, gestures, and/or attempts.
C. Document mental health professionals (e.g., psychiatrists, psychologists, counselors, etc.) who treated the decedent.
D. Document family mental health history.

Summary: Knowledge of the mental health history allows the investigator to evaluate properly the decedent's state of mind and contributes to the determination of cause, manner, and circumstances of death.

5. Document Social History

Principle: Social history includes marital, family, sexual, educational, employment, and financial information. Daily routines, habits and activities, and friends and associates of the decedent help in developing the decedent's profile. This information will aid in establishing the cause, manner, and circumstances of death.

Authorization: Medical Examiner/Coroner Official Office Policy Manual; State or Federal Statutory Authority.

Policy: The investigator shall obtain social history information from sources familiar with the decedent.

Procedure: When collecting relevant social history information, the investigator should:

A. Document marital/domestic history.
B. Document family history (similar deaths, significant dates).
C. Document sexual history.
D. Document employment history.
E. Document financial history.
F. Document daily routines, habits, and activities.
G. Document relationships, friends, and associates.
H. Document religious, ethnic, or other pertinent information (e.g., religious objection to autopsy).
I. Document educational background.
J. Document criminal history.

Summary: Information from sources familiar with the decedent pertaining to the decedent's social history assists in determining cause, manner, and circumstances of death.

Section F: Completing the Scene Investigation

1. Maintain Jurisdiction over the Body

   Principle: Maintaining jurisdiction over the body allows the investigator to protect the chain of custody as the body is transported from the scene for autopsy, specimen collection, or storage.

   Authorization: Medical Examiner/Coroner Official Office Policy Manual; State or Federal Statutory Authority.

   Policy: The investigator shall maintain jurisdiction of the body by arranging for the body to be transported for autopsy, specimen collection, or storage by secure conveyance.

   Procedure: When maintaining jurisdiction over the body, the investigator should:

   A. Arrange for, and document, secure transportation of the body to a medical or autopsy facility for further examination or storage.
   B. Coordinate and document procedures to be performed when the body is received at the facility.

   Summary: By providing documented secure transportation of the body from the scene to an authorized receiving facility, the investigator maintains jurisdiction and protects chain of custody of the body.

2. Release Jurisdiction of the Body

   Principle: Prior to releasing jurisdiction of the body to an authorized receiving agent or funeral director, it is necessary to determine the person responsible for certification of the death. Information to complete the death certificate includes demographic information and the date, time, and location of death.

   Authorization: Medical Examiner/Coroner Official Office Policy Manual; State or Federal Statutory Authority.

   Policy: The investigator shall obtain sufficient data to enable completion of the death certificate and release of jurisdiction over the body.

   Procedure: When releasing jurisdiction over the body, the investigator should:

   A. Determine who will sign the death certificate (name, agency, etc.).
   B. Confirm the date, time, and location of death.
   C. Collect, when appropriate, blood, vitreous fluid, and other evidence prior to release of the body from the scene.
   D. Document and arrange with the authorized receiving agent to reconcile all death certificate information.
   E. Release the body to a funeral director or other authorized receiving agent.

Summary: The investigator releases jurisdiction only after determining who will sign the death certificate; documenting the date, time, and location of death; collecting appropriate specimens; and releasing the body to the funeral director or other authorized receiving agent.

3. Perform Exit Procedures

Principle: Bringing closure to the scene investigation ensures that important evidence has been collected and the scene has been processed. In addition, a systematic review of the scene ensures that artifacts or equipment are not inadvertently left behind (e.g., used disposable gloves, paramedical debris, film wrappers, etc.) and any dangerous materials or conditions have been reported.

Authorization: Medical Examiner/Coroner Official Office Policy Manual; State or Federal Statutory Authority.

Policy: At the conclusion of the scene investigation, the investigator shall conduct a post-investigative "walk through" and ensure the scene investigation is complete.

Procedure: When performing exit procedures, the investigator should:

A. Identify, inventory, and remove all evidence collected at the scene.
B. Remove all personal equipment and materials from the scene.
C. Report and document any dangerous materials or conditions.

Summary: Conducting a scene "walk through" upon exit ensures that all evidence has been collected, that materials are not inadvertently left behind, and that any dangerous materials or conditions have been reported to the proper entities.

4. Assist the Family

   Principle: The investigator provides the family with a timetable so they can arrange for final disposition and provides information on available community and professional resources that may assist the family.

   Authorization: Medical Examiner/Coroner Official Office Policy Manual; State or Federal Statutory Authority.

   Policy: The investigator shall offer the decedent's family information regarding available community and professional resources.

   Procedure: When the investigator is assisting the family, it is important to:

   A. Inform the family if an autopsy is required.
   B. Inform the family of available support services (e.g., victim assistance, police, social services, etc.).
   C. Inform the family of appropriate agencies to contact with questions (medical examiner/coroner offices, law enforcement, SIDS support group, etc.).

D. Ensure family is not left alone with body (if circumstances warrant).
E. Inform the family of approximate body release timetable.
F. Inform the family of information release timetable (toxicology, autopsy results, etc., as required).
G. Inform the family of available reports, including cost, if any.

Summary: The interaction with the family allows the investigator to assist and direct them to appropriate resources. It is essential that families be given a timetable of events so that they can make necessary arrangements. In addition, the investigator needs to make them aware of what and when information will be available.

Notes: "It is important to note that even the use of the word 'system' to describe a process that encompasses more than 3,000 individual jurisdictions is a misnomer." Hansen, M., "Body of Evidence," American Bar Association Journal (June 1995).

Chapter 8

AVIATION ACCIDENT INVESTIGATIONS

This chapter will give a brief overview as 99% of investigations will come from the NTSB. Having a familiarity with the concept and techniques will help you determine if this is a good career move for you.

The Forensic Services Program is the focal point of National Weather Service (NWS) support for transportation accident investigations and potential subsequent litigation. Much of the work is related to aviation accidents, but a considerable amount of effort is related to marine and other surface accidents. The program activity includes:

Ensuring that high-level National Weather Service and National Oceanic and Atmospheric Administration (NOAA) management are provided timely notification, and kept informed of transportation accidents of unusual interest.

Upon request, assisting the *National Transportation Safety Board (NTSB)* in its investigation of transportation accidents.

Representing the National Weather Service at *NTSB Public Hearings* and preparing NWS personnel designated as witnesses to testify at Public Hearings.

Responding to NTSB Safety Recommendations addressed to the NWS evolving from transportation accident investigations.

Maintaining an awareness of all investigative and/or legal activity evolving from weather-related transportation and pipeline accidents.

Providing technical advice to Government attorneys defending suits against the Government evolving from transportation accidents.

Assisting Government attorneys in preparing NWS employees for testimony at *legal proceedings*.

Responding to legal production, Freedom of Information Act (FOIA), and mission information requests for *weather records documentation*.

Ensuring the NWS records retention program meets agency litigation-support requirements.

NTSB Field Investigation

The NTSB is responsible for the investigation and determination of probable cause of all civil aircraft accidents, certain highway and railroad accidents, pipeline accidents in which there is a fatality or substantial property damage, and certain major marine casualties. Many of these transportation accidents are relatively low key, with the investigations handled locally by NTSB regional staff. Some are delegated to the Federal Aviation Administration (FAA).

However, after any fatal accident involving an air carrier, NTSB's national headquarters will conduct the field investigation, to determine the facts, conditions, and circumstances related to the event. A "go-team," headed by the NTSB investigator-in-charge is at the accident scene within 24 hours of the accident. A field investigation will last anywhere from one to three weeks, depending on the scope of the accident in question. When the accident is suspected to be weather-related, the "go-team" will include an NTSB meteorologist and may include a representative from the NWS - normally a member of the forensic services staff. As a member of the weather group, the NWS representative assists in gathering appropriate "work" documents and interviewing NWS personnel, and participates in the development of the written factual record of the on-scene phase of the accident. In cases where there is no NWS representative on the "go-team," the NTSB meteorologist will contact a designated NWS regional official, usually the regional aviation meteorologist (RAM), or equivalent, for assistance. The RAM will act as a liaison between the NTSB and NWS offices and personnel. A list of NWS liaisons is maintained and routinely updated by the forensic services program and passed on to the NTSB.

NTSB Public Hearing

The NTSB, at its discretion, may convene a public hearing as a part of its investigation into a transportation accident for the purpose of creating a public record of the facts, conditions, and circumstances relating to the accident. A public hearing usually occurs within two to three months after an accident. It normally lasts for a week or less (but may be considerably longer - the hearing evolving from a Trans World Airlines crash near Berryville, Virginia, in 1974, lasted for four weeks)!

The key players at NTSB Hearings are the NTSB Technical Panel, the Spokespersons for the various designated parties, and the witnesses. The NWS is normally a designated party in weather-related cases where the testimony of NWS employees is required. Commonly, NWS witness roster will include the duty surface weather and radar observers, the duty NWS office forecaster, and the duty center weather service unit meteorologist. Most of the questions addressed to a participating witness will originate from a member of the technical panel. Where NWS witnesses are concerned, this member will be the NTSB meteorologist assigned to the case. However, all of the spokespersons of the various designated parties, some of which may be preparing for subsequent litigation, are permitted to ask questions.

NTSB Aircraft Accident Report

Upon completion of its investigative activities, the NTSB publishes the Aircraft Accident Report. With respect to carrier accidents, this will occur anywhere from one to two years after the accident. The report contains detailed factual information as well as an extensive analysis of the factors and circumstances surrounding the accident question. It also itemizes findings of the NTSB and assigns probable and contributing causes. The latter is the only part of the report which cannot be used in litigation. It is more than a coincidence that potential legal activity is normally deferred until after the NTSB has published its report.

The Aircraft Accident Report will commonly include multiple safety recommendations relating to issues surrounding the accident in question. These will be addressed to the appropriate government agencies, as well as civil parties and organizations. Based on its investigations and findings, the NTSB will identify deficiencies and oversights, and make recommendations for remedial or corrective action, so as to prevent a similar reoccurrence in the future. The forensic services program is the NWS focal point for responding to safety recommendations relating the aviation weather operations and services.

Litigation Process

Overview

Legal activity in cases involving the NWS, from the time that administrative claims are presented to the agency, normally lasts for up to five years - although cases have gone on as long as ten years. During this period, forensic services staff may be required to produce hundreds of operational and administrative documents related (or determined to be related) to the case, and spend many hours dedicated to providing technical advice to government attorneys and assisting in preparing NWS employees for testimony. It is noteworthy that the 46 cases producing administrative claims against the NWS during the period from 1980 through 1998 had total monetary attachments of almost 1.6 billion dollars! The enormous amount of time and resource investment in defense against these inflated claims is made not only to conserve U.S. treasury funds, but of equal importance, to prevent a court ruling which may adversely affect future NWS operations and/or its employees to exercise their best professional judgment.

Documentation Case Files

The Forensic Services Program establishes and maintains about 250–400 each year.

Each case represents a request or often several requests for weather information.

Response to weather documentation requests includes the retrieval, certification, and production of historic aviation and non-aviation (e.g. marine, surface, hydrology) records

Requests for weather documentation may come from private citizens; weather consultants; attorneys; and other federal, state, or local government agencies

Requests for weather data may be simple and straight forward while others may take a considerable effort and cooperation from many sources to fulfill

- By most of the case files handled by the forensic services are related to aviation accidents; however, a large number are related to marine and other surface accidents
- The public accounts for most of the customer requests. Primary Government customers are the NTSB, the FAA and the DOJ. Other miscellaneous customers include the US Coast Guard, offices within the Department of Commerce, and state and local law enforcement agencies.

Past weather records can be obtained in 2 ways:

For all non-certified records or certified climatological records (including radar images, satellite photos, surface analysis, buoy or ship reports), please contact:

National Climatological Data Center
151 Patton Avenue
Asheville, NC 28801-5001
(828) 271-4800

For CERTIFIED FORECASTS (or any requests that include the need for certified forecasts), the requests MUST be in writing, either via mail or fax. Please contact:

NWS Headquarters
ATTN: W/OS52
1325 East-West Highway
Silver Spring, MD 20910

Facsimile: (301) 713-1598 (ATTN: Wayne Presnell or Alfreda Walters)
Voice: (301) 713-0090 x147 or x176

(Courtesy: http://www.nws.noaa.gov/om/forensic.shtml)

Microbursts are a phenomena that has been the cause for many aviation accidents. Fernando Caracena and Dr. Fujita set out to show a classical situation of a microburst.

Collaborating with Dr. Fujita

Fernando Caracena, NOAA, Forecast Systems Laboratory, Boulder, Colorado

In 1976 Don Beran (of NOAA's Wave Propagation Laboratory) suggested that I write an article about the crash of Continental flight 426 as a contribution to a special windshear issue of the Bulletin of the American Meteorological Society (BAMS), which he was editing—perhaps collaborating with Dr. Fujita who was also preparing a BAMS article on the crash of Eastern Flight 66. At Fujita's invitation, I flew up for over a week's work with him at his laboratory at the University of Chicago. By then, Allegheny Flight 121 had crashed while attempting to land at Philadelphia International Airport on the afternoon of 23 June 1976 as it emerged from a "wall of water." For over a week, we worked side by side, analyzing the three accidents using our combined data sets.

Three aircraft taking off within minutes of each other experienced increasing windshear from a developing microburst.

Fujita's weather analysis technique

Fujita made mesoscale weather analyses look like a leisurely drive down a country road on a fine autumn day. He told me, "The hardest part is collecting all the data." For Fujita, the rest was easy. . . . In the end, Fujita wove a four-dimensional weather analysis, superposed on radar imagery that covered the accident time and location.

A small circular radar echo formed west of the airport 4 min before Continental 426 took off. Six minutes later, the echo had grown into a large spearhead that covered the north end of the airport. Meanwhile, the microburst was impacting on runway 35L.

I was gratified to see that Fujita's analysis agreed with mine. But, his was much richer in detail. The valuable weather analysis techniques that I learned by collaborating with Dr. Fujita would serve me well in the future.

Historical importance of the three weather-related accidents

The years, 1975 and 1976 were landmarks in air safety history. First, Fujita identified the microburst as a threat to aviation. Second, our weather analysis of the Denver accident documented that a microburst was the cause of the crash. Finally, a third crash in a short span of time made clear the mounting threat of microbursts to aviation. The microburst was a beast that ate airplanes. Meanwhile, some aircraft weather experts testified that this beast could not exit. But, Fujita continued to compile a growing list of crashes involving microbursts that grew to 8 by 1985.

Greg MacMaster

Wet and Dry Microbursts

(Fernando Caracena NOAA, Forecast Systems Laboratory, Boulder, Colorado)

Wet and dry microbursts-operational definitions

The field experiments of the 1980s and 90s (NIMROD, JAWS, MIST and others) revealed microburst wind gusts striking a network site with measurable rain, and others, without. Two of the three accidents of 1975 and 1976 involved heavy rain and another, only vanishing sprinkles.

NIMROD (Northern Illinois Meteorological Research on Downburst, Chicago, 1978–79),

JAWS (Joint Airport Weather Studies, Denver, 1982–86)

MIST (Microburst and Severe Thunderstorm project, Huntsville, Ala., 1986–88).

Dry storms can exist in environments having high lapse rates. Braham (1985) and W. R. Krumm (1954) describe a dry, high-based thunderstorm and its environment common in the western United States, which produces lighting, almost no rain, but lots of strong, sometimes damaging wind.

Brown et al. (1982) identify an even more extreme, dry storm environment (Fig. 2 a), in which the dry static stability almost vanishes and just enough moisture stratification remains to support cumulonimbi without lightning, but with just enough precipitation to drive severe downdrafts. Weak updrafts and cold condensation temperatures grow precipitation mostly as lightly rimed ice crystals, which falling through the melting level, become a spray of small raindrops that evaporate rapidly.

A local sounding taken at the Florida Area Cumulus Experiment three hours before a wet microburst struck this site reveals that the source of the strong downdraft was an elevated dry layer containing air potentially cold enough to inject a cool downdraft into the precipitation column at a high level from where it could accelerate to the surface.

Microburst information Courtesy:

Fernando Caracena NOAA, Forecast Systems Laboratory, Boulder, Colorado PhD, Physicist

Greg MacMaster

Mesoscale Applications Group
Forecast Research Division
Forecast Systems Laborator
NOAA

Thunderstorms, icing and turbulence are a determent to aviation. If thunderstorms are forecast, severe icing and severe turbulence is implied and flight should be restricted into these storms. However, there are some cloud types that suggest severe up and downdrafts without showing the characteristics. I'm referring to the Cb cloud. A cumulonimbus cloud (without the anvil) has the same characteristics of a Cb with an anvil. Strong up and down drafts, severe turbulence and of course, severe icing are hazards pilots try to stay away from.

Forecasting Icing Using the -8D Rule

1. At significant levels on the Skew-T, determine temperature and dew-point spread. The result is the "D" value.
2. Multiply the "D" value by -8 and plot at the significant level.
3. Connect the plotted points by drawing a straight line from point to point.
4. Where line crosses right of temperature trace, clouds are supersaturated with respect to ice and icing is possible.

Icing Intensity (White Sands Missile Range)

L = (1/2.87) * (w_o - w) * (P/t). W_o and W is the mixing ratio at cloud base and at level in question. P is the millibar pressure level in question and the T is the temperature in Kelvin (273°C)

Other Rules for Forecasting Moderate or Greater Clear Air Turbulence

1. Whenever winds aloft, vertical shear, equals or exceeds 30 knots per 5000 feet.
2. For jet stream stronger than 110 knots at core, particularly in sloping tropopause above core.

3. If 300mb 20 knot isotachs are less than 60 nautical miles apart.
4. In entrance zone between two confluent jet cores in region where cores are 3–5° latitude or less apart.
5. In sharp, V-shaped troughs which slope rapidly with elevation below 300mb.
6. In "neck" of cut-off lows.
7. To right of strong anticyclonically curved jet streams.
8. Vertical extent is no higher than 2000 feet above tropopause and within layers described above.
9. Strong horizontal temperature gradients (10°C/100 miles) signify strong vertical wind shear and turbulence.

Turbulence Chart for Aviation (According to Aircraft Category)

Category I	Category II	Category III	Category IV
N	N	N	N
(L)	N	N	N
Light	(L)	N	N
L-(M)	L	(L)	N
M	L-(M)	L	(L)
M-(S)	M	L-(M)	L
S	M-(S)	M	L-(M)
S-(X)	S	M-(S)	M
X	S-(X)	S	M-(S)
X	X	S-(X)	S
X	X	X	S-(X)
X	X	X	X

L = Light M = Moderate S = Severe

This chart displays the severity of turbulence according to the type of aircraft that fits within that category. Category I aircraft depicts lighter aircraft (Cessna 152 for example) all the way to category IV which includes the heavier aircraft, jet fighters, larger transport carriers.

SEVERE THUNDERSTORMS AND TORNADOES

Temperature The most vigorous storms occur in air having a morning subsidence inversion, a characteristic associated with wind-shear.

Moisture The strongest storms require drier air above the ridge or a marked dry source to the windward side in a position to intrude into or over the moist ridge.

Winds Strong mid-level winds are required with sharp horizontal wind-shears. Moderate to strong the low level flow is required for major storm family type outbreaks. The intersection of the maximum low-level winds with a warm front or old squall line is a frequent development area. It is extremely desirable for the mid-level and low-level jets to intersect, since the tornado activity major axis is frequently determined by the location and movement of this intersection.

Vorticity Strong PVA interacting with a dryline leads to a much higher probability of igniting intense to severe thunderstorms.

Stability of the air column depends largely on the low-level moisture and temperature advection.

Upper-level Jet Cyclogenesis can result when there is a dynamic coupling of the low-level jet's right-front with the upper-level jet's left-front quadrant. There is a significant correlation between severe storm development when a southerly low-level jet is bisected at an angle of 30° or more.

Low-level Jet The highest upward lifting rate (enhancing low-level warm advection) is associated with the nose and the right-front quadrant of the low-level jet.

Low-level Moisture is best determined from the sounding's analysis of the lowest 3,000 feet using the moist layer's mean dew point. Or, take the 850mb and surface dew point temperature averages.

850mb Temperature Pattern The maximum temperature ridge is located to the west or southwest (upwind) of the low-level moisture ridge. It is extremely important to locate the highest rate of warm advection area. The low-level thermal ridge is associated the with highest lapse rates but is low in moisture, while the low-level moisture axis downstream has marginal conditional instability in levels above it. When these regions overlap, the potential for intense convection is greatest. The vertical totals index helps to diagnose steep lapse rate regions.

700mb No-change Line connects points of no advective temperature changes. This line separates warm air advection areas from cold air advection areas. It is coincident with the 850mb warm ridge. It has been observed that the advancement of the no-change line's leading edge into a position ahead of a pronounced mid-level trough is usually associated with deepening or intensification of surface low centers.

700mb Dry Air Intrusion Consider 50 percent relative humidity, dew point less than 0°C, or a temperature/dew point spread of more than 6°C at 700mb as dry. The most favorable relationship between the low-level and 700mb flow occurs when the winds veer with height.

Surface Pressure below 1013mb, and never above 1018mb.

Surface Dew Point The incidence of tornadoes is infrequent at dew-point temperatures below 13⁰C.

Twelve-hour Surface Pressure Falls reliably indicate major changes or trends in the surface pressure pattern and reflect important changes aloft. In the most productive situations, it appears that widespread pressure falls are less desirable than a pattern of more concentrated falls.

Winds - Veering and Backing

The wind veers when it changes direction clockwise. Example: The surface wind is blowing from 270°T. At 2000 feet it is blowing from 280°T. It has changed in a right-hand, or clockwise, direction.

The wind backs when it changes direction anti-clockwise. Example: The wind direction at 2000 feet is 090°T. at 3000 feet 0850T. It is changing in a left-hand, or anti-clockwise, direction.

In a descent from several thousand feet above the ground to ground level, the wind will usually be found to back and also decrease in velocity, as the effect of surface friction becomes apparent. In a climb from the surface to several thousand feet AGL, the wind will veer and increase.

At night, surface friction decreases as surface cooling reduces the eddy motion of the air. Surface winds will back and decrease. During the day, as surface friction intensifies, the surface winds will veer and increase.

Winds - Wind Shear

Wind shear is the sudden tearing or shearing effect encountered along the edge of a zone in which there is a violent change in wind speed or direction. It can exist in a horizontal or vertical direction and produces churning motions and consequently turbulence. Under some conditions, wind direction changes of as much as 180 degrees and speed changes of as much as 80 knots have been measured.

The effect on airplane performance of encountering wind shear derives from the fact that the wind can change much faster than the airplane mass can be accelerated or decelerated. Severe wind shears can impose penalties on an airplane's performance that are beyond its capabilities to compensate, especially during the critical landing and take-off phase of flight.

Winds - At Cruising Flight

In cruising flight, wind shear will likely be encountered in the transition zone between the pressure gradient wind and the distorted local winds at the lower levels. It will also be encountered when climbing or descending through a temperature inversion and when passing through a frontal surface. Wind shear is also associated with the jet stream (see below). Airplanes encountering wind shear may experience a succession of updrafts and downdrafts, reductions or gains in headwind, or wind shifts that disrupt the established flight path. It is not usually a major problem because altitude and airspeed margins will be adequate to counteract the shear's adverse effects. On occasion, however, the wind shear may be severe enough to cause an abrupt increase in load factor, which might stall the airplane or inflict structural damage.

Winds - Near the Ground

Wind shear, encountered near the ground, is more serious and potentially very dangerous. There are four common sources of low level wind shear: thunderstorms, frontal activity, temperature inversions and strong surface winds passing around natural or manmade obstacles.

Frontal wind shear. Wind shear is usually a problem only in fronts with steep wind gradients. If the temperature difference across the front at the surface is 5°C or more and if the front is moving at a speed of about 30 knots or more, wind shear is likely to be present. Frontal wind shear is a phenomenon associated with fast moving cold fronts but can be present in warm fronts as well.

As a pilot, flying in weather that is pushing the limits of my ability creates stress and alters my state of mind. Needless to say, causes of

Greg MacMaster

accidents can point to one variable or be innumerable. In this next case, we find that our stress level increases when the weather worsens, but how much?

AWARENESS OF STRESSES ASSOCIATED WITH APPROACHES AND LANDINGS UNDER MARGINAL CONDITIONS

di J.S. Clauzel Flight Safety-DACO-1981

While the total numbers of fatal accidents during approach and landing are decreasing, those occurring in marginal weather conditions appear to be increasing. Accordingly, this paper presents a comprehensive examination of the factors that contribute to this type of accident; namely, weather conditions, aircraft performance, airport characteristics, and flight crew human factors. Effects of control inputs as related to stall margins, visual illusions, and other phenomena associated with marginal weather conditions are addressed. Recommendations are offered on crew conditioning, with the position token that, during an approach, when there is sufficient doubt in the minds of the flight crew, a go-around should be viewed as a mark of good judgment. Further, it is recommended that each airline carefully review this area of operations with respect to training and policy.

Introduction

The Operational Flight Safety group at Douglas Aircraft Company has been investigating several recent accidents, incidents, and near accidents which have occurred during landing attempts under marginal approach conditions. One of these was a major disaster, and the potential for more is there. None was associated with any aircraft system malfunction. Three of the incidents will be presented for purposes of illustration, although these are only a few of the many available.

1. The aircraft made a hard landing 45 feet short of the runway. Weather was rain and fog. Flight recorder peaked at 3.5g.
2. Upon landing, all tires on the main landing gear blew out. Weather conditions were reported as marginal with light rain. The following items were damaged: left-hand main landing gear strut door, inboard portion of flap trailing edge, and both main landing gears.
3. On landing, the aircraft touched down and departed the runway. Both landing gears went off the runway and hit several runway lights. The aircraft traveled approximately 2000 feet on the ground, went across a service road, took off again, and made a go-around and a normal landing. Weather was marginal. Damage was sustained on the undercarriage.

It is apparent that common factors in each of these occurrences were abnormal touchdowns and the decision to attempt to land the airplane under marginal approach conditions. Although extensive research has been conducted on this subject and numerous papers written, it is apparent that we in the aircraft safety business still have much work to do. Based upon statistics which show that almost 50 percent of all air carrier accidents occur during approach and landing and that many more are very close calls, we must conclude that we have not yet reached the person in the cockpit in an effective way. As always, the key to an effective safety effort is cockpit awareness of the hazard combined with a concise definition of the problem and clear guidelines for dealing with it. While it is true that many of the most important decisions in the cockpit must be made in the proverbial split second, the decision arrived at is more likely to be the right one if the crew is properly prepared for it ahead of time.

In most cases, an accident, incident, or near accident during approach or landing occurs under marginal conditions because the flight crew elects to continue the approach even though things are not quite right, obviously hoping to catch up. It is the author's conviction that missed approach procedures are not being utilized when, in fact, weather, airport, aircraft, flight crew conditions, or a combination of these factors dictate that they should be utilized. Seldom is any single factor in aircraft operation the sole cause of anything. Rather, the entire complex, interwoven pattern must be considered. An event which occurs minutes or even hours before a landing may, in the final analysis, make the landing unsafe. *This paper will review the various facets of marginal approach conditions, including the weather, the airport, the aircraft, and the flight crew, with the view that flight crews must be made aware of possible pitfalls and be encouraged to execute a missed approach when it is prudent to do so, rather than to unnecessarily risk a hazardous landing.*

Statistical review of approach and landing fatal accidents

To assess the significance of the effects of weather on fatal accidents in approaches and landings, statistics from Flight International from 1975 through 1980 were reviewed. During this six-year period, there were 136 fatal accidents, of which 51 (37.5 percent) occurred during the approach and landing phase of flight. Of the 51 fatal approach and landing accidents, 10 were listed as being related to adverse weather conditions. These 10 accidents thus represent approximately 7.4 percent of all fatal accidents and close to 20 percent of all fatal approach and landing accidents. From the author's own compilations, we know that there were many more accidents and incidents in this category that were not fatal, plus even more very close calls.

Figure 1 shows the relationship between all fatal accidents and those fatal accidents occurring during approaches and landings. The approach and landing accidents are separated into weather and non weather-related contributory causes. The figure represents a smoothing of the data by using best-fit log-normal regression lines to enable an interpretation of the general trends of the data. This reveals some significant trends in fatal accidents and, more specifically, in approach and landing accidents. First, it appears that during the past six years the total number of accidents along with the approach and landing subset are decreasing. The number of non-weather-related approach and landing fatal accidents also appear to be decreasing, but at a greater rate. However, we now find that fatal approach and landing accidents where weather is a factor exhibit an opposite trend and that rate is, in fact, increasing.

In the analysis of marginal weather-related approach and landing accidents, there are several major factors which should be considered.

The first concerns the specifics of the weather conditions at the time of the accident; the ceiling, visibility, cloud cover, fog, wind shear, etc. The second factor is the aircraft. Were there any system malfunctions which altered the flight characteristics? Were there items on the minimum equipment list (MEL) which might increase workload? Were there basic flight or control characteristics which might influence aircraft performance in marginal conditions? The third factor concerns the flight crew. How well prepared were they to cope with the unique conditions they encountered? Did they have sufficient sleep the night before? Are they on medication, either prescription or non-prescription? What kind of marital, financial, or other social problems do they have? Then there is the airport at which the landing is to be made.

FIGURE 1 WORLDWIDE FATAL ACCIDENT TRENDS - 1975 TO 1980

Runway length, width, grooving, approach aids, lighting, and other items each play an important part in an approach and landing under marginal conditions. These factors will now be analyzed in more detail.

Greg MacMaster

What constitutes marginal approach conditions?

The effects of adverse weather conditions confronting an aircraft have been well covered in available literature, and much has been written about frontal activity along with wind shear, rain, fog, scud, and the like. While these phenomena need not be described here, it should be recalled that weather can make the aircraft's environment not only hostile, but deadly. On one hand, weather can rapidly affect the entire air mass through which the airplane is moving, thereby changing the dynamic responses of the airplane; and, at the same time, it can produce marginal visibility, and further change the effective dynamics of the airplane through delays or misinterpretations in pilot "feedback" especially during the critical maneuvering to land after sighting the runway during an instrument approach. And even with today's sophisticated weather satellites and weather measuring and advisory systems, the crew must still make the final decision of whether to land or go around.

There is little pilots can do about rapidly changing air masses, with respect to wind shear, for example, aside from avoiding the condition or carrying sufficient speed (energy) so that when the shear is encountered, with application of proper technique, the aircraft will continue to fly. Similarly, there is little a pilot can do about reduced visibility except to cope with it as best he can. Unfortunately, there is no practical method of measuring the most important parameter "slant visual range," which is the pilot's maximum visibility over the aircraft nose toward a ground-based object. This can be deduced from other parameters that are measured such as the prevailing visibility at the airport, ceiling, and runway visual range. However, this leaves much to be desired, especially in scud or other non-homogeneous visibility conditions near the "decision height," or minimum descent altitude. The runway could be in sight at the decision height, but the visual situation could change drastically in a moment so that when a pilot breaks out of overcast conditions a second time, he could find the visual approach slope indicator (VASI) reading in the red zone, and shortly thereafter hear the sounds of impact with the tops of trees. The moral of this story is: when in doubt, execute a missed approach.

Accurate, up-to-date airport weather information is vital for safety during landing operations.

The airplane's contribution to a marginal approach is, in some cases, obvious. Yet, in some cases, it is not so obvious—such as when dispatch with a particular system inoperative may later contribute adversely to the marginal approach. The basic avionics configuration of a particular model alone can be restrictive if the airplane's instrument flight capability is insufficient to provide the crew with the proper tools for the prevailing conditions.

Crew members' knowledge of the airplane's performance capabilities can also be an important factor, depending on their knowledge of the hazardous condition. In a severe wind shear, for example, we recommend that during a rapid decrease in headwind or increase in tailwind, which tends to decrease indicated airspeed, thrust and pitch attitude be immediately increased to maintain an acceptable airspeed and flight path. Power should be immediately advanced to the maximum allowable setting, and a missed approach should be initiated when this type of an encounter occurs at low altitude. Airspeed should be traded down to the stick shaker speed (but not below), if necessary, to prevent ground impact.

Turbulence is another reason for selecting higher final approach speeds for better controllability. Most airline operators require their pilots to modify their approach speeds so that in conditions of gustiness or turbulence, speed increments are added. Pilots are generally told to apply these increments above 1.3 V_s and to fly at the higher speed during the approach in order to minimize the risk of stall or loss of control as a result of turbulence, gust, or wind shear. The effect of control inputs during extensive maneuvering close to flare speed especially during turbulence will be dealt with in more detail later.

In addition to turbulence, wake vortex—or more accurately, tip vortex—during an approach can greatly influence the approach path. A twin-jet passenger-liner on a training flight rolled completely over and crashed, fatally injuring four, as the result of an encounter with a wake vortex generated by a wide-bodied aircraft at Forth Worth. The mechanics of wake vortex generation are fairly well documented, and adequate protection should be afforded by proper separation of aircraft. Another form of turbulence, sometimes found very close to the ground, is generated by vortices shed off large buildings near runways. At some airports, new, large hangars have been built for wide-bodied jets, and even a light breeze can cause disturbing vortices to be shed into the aircraft's flight path. Pilots should be aware of the condition by becoming familiar with airport layouts where prevailing winds cause the problem, and determine how much more speed to carry on final approach in order to compensate for it.

Another factor in marginal approach conditions is the so-called visual illusion. This can adversely affect the pilot's perception of the approach scene. The touchdown distance can be greatly influenced by optical illusions which can take many insidious forms, especially on a visual approach to a sloped runway. The pilot approaching a runway sloped up-hill or even a level runway preceded by up-hill terrain, will have an above glide-path illusion that makes him think he is high—the normal glide slope will appear too steep—but he may, in fact, be low, flying a flat approach, and thus be exposed to a short landing. The opposite effect will occur during an approach to a down-hill runway, or over down-sloping terrain, and the pilot will tend to fly a steeper approach and thus be exposed to landing long or even overshooting the runway. Hazards near the runway may produce detrimental psychological effects if the runway is short, and if the pilot attempts to touch down close to the threshold, the danger could be increased.

Rain on the windshield can interfere disastrously with a pilot's assessment of his progress toward touchdown. Rain causes objects to be obscured by an overall blurring effect. Light dispersion through the layers of water on the windshield can cause the horizon to appear to the pilot as a gradient blur. Pilot assessment of touchdown is also adversely affected by runway lighting, color contrast, and other characteristics.

Thus, approaches to brightly lighted runways at night have resulted in touchdowns short of the runway. The effect is greatly increased in clear air or when the approach zone is not lit. Irregularities in runway surfaces can cause a runway to appear much shorter, especially if the pilot loses sight of the far end of the runway because of a ground rise between the aircraft and the far end. When the color of the runway approximates the color of the surrounding terrain, problems in depth perception occur.

Examples are the snow-covered runway and the dimly lit runway. But even less obvious conditions present severe problems in pilot perception, resulting in overshoots and undershoots.

The experienced pilot has trained his eye to subtend an arc of some 2-1/2 to 3 degrees during an approach, the angle related to the horizon which determines the proper touchdown aiming point. But the pilot will automatically adjust this angle downward, applying it from the far end of the visual segment; thus, in the absence of other guidance or cues, insidiously aiming himself short whenever only part of the runway is in view.

One of the most consistent statistics that emerges from extensive studies of causative factors in aircraft accidents is that 50 to 70 percent can be attributed to errors made by the flight crew. A typical breakdown is shown in Figure 2.

PROBABLE CAUSE	NO. OF ACC	PERCENT OF TOTAL ACCIDENTS
COCKPIT CREW	249	64.5%
WEATHER	13	3.4%
AIRPLANE	82	21.2%
MAINTENANCE	11	2.8%
AIRPORT	13	3.4%
MISCELLANEA	18	4.7%
TOTAL WITH KNOWN CAUSES	386	

FIGURE 2 CAUSE FACTORS - ALL TYPES OF ACCIDENTS (TYPICAL)

We already know that of all accidents and incidents, approximately 50 percent are in the approach and landing phase. In this classification, slightly less than 70 percent are relegated to human error with the following cited as major causal factors.

1. Failed to follow approved procedures.

2. Improper operation under instrument flight rules.
3. Misjudged speed/altitude.
4. Spatial disorientation.
5. Inadequate supervision of flight.

The term "human error" is, however, misleading in that the "erring" is usually equated to failure to properly perform a given task. If the demands of the task exceed the pilot's capacity, then it can be said that he has not erred.

In reality, the performance has been within human limitations. The actual error may be looked upon in two ways:

1. The system has failed to recognize the pilot's limitations
2. The pilot has failed to recognize his own limitations.

In either case, steps must be taken to either increase the pilot's capacity by training or reduce the demands, for example, by executing a missed approach in order to avoid an accident-prone situation. The former is a long-term solution and the latter an immediate one. In general, aviation can be considered as a high-stress occupation for flight crews. They are involved in critical tasks in which human errors can be catastrophic and they are subjected to adverse environmental factors such as low humidity, widely varying temperatures, and noise. In addition, excessive social problems may result from family separation, irregular work periods, and schedules that amount to an almost permanent stand-by status. The effects of constantly having to resynchronize their body rhythms, which have been disturbed by time-zone changes and irregular work/rest patterns, and the sleep disturbance and deprivation which results from these living patterns must also be taken into consideration. This is especially true at the end of a long day, as shown in Figure 3, when the very high stress level demands of an approach and landing in adverse weather conditions may often be experienced.

FIGURE 3 - PERFORMANCE DEMANDS AS RELATED TO PERFORMANCE ABILITY

Obviously, a large number of factors associated with the physiological and psychological makeup of the flight crew tend to degrade or limit human performance capability.

Table 1 lists a number of these which, when taken in the context of marginal weather conditions, tend to advance the incident toward the accident.

While it is not the purpose of this paper to detail all of the stressors placed upon flight crews, the manifestations of several will be explained to provide some understanding of how these stress factors affect performance at crucial phases of flight.

For example, fatigue, which is difficult to measure accurately or quantify, manifests itself in symptoms which can be clearly seen and felt, such as:

1. Decreased attention.
2. Slowed and impaired perception.
3. Impaired thinking.
4. Decreased motivation.

5. Decreased physical and mental performance.

TABLE 1

PHYSIOLOGICAL AND PSYCHOLOGICAL FACTORS WHICH LIMIT HUMAN PERFORMANCE

1. PREEXISTING ILLNESS/DEFECT	9. LACK OF FAMILY SUPPORT
2. GENERAL PHYSICAL CONDITION	10. ANGER
3. NUTRITIONAL STATE (HUNGER)	11. FRUSTRATION
4. DRUGS (SELF-MEDICATION)	12. WORRY
5. FATIGUE (LOSS OF SLEEP)	13. OVERSENSITIVITY
6. CIRCADIAN RHYTHM (TIME ZONE CHANGES)	14. GUILT
7. AGE	15. MEMORIES OF PAST INCIDENTS
8. TASK PROFICIENCY	16. HANGOVER

Fatigue has a tendency to lower a person's assessment of the acceptable level of his performance. It lowers the level of quality which he or she believes should be attained in task performance.

Another stress factor to be considered, which has effects similar to fatigue, is the use of various medications (prescribed or over-the-counter) by flight crews in the treatment of the common cold or to induce sleep. It should be realized that a large number of these medications affect the crews' task performance in a detrimental way and have predictable side and after effects. Some can cause allergic reactions when used in combination with alcohol.

Others can cause very serious side effects long after ingestion. Use of over-the-counter antihistamines to alleviate symptoms of the common cold should also be approached with care. The case is cited of a pilot taking Sudafed, a nasal decongestant, which resulted in postural dizziness associated with "graying of vision." This situation re-emphasizes the potential hazard of self-medication for upper respiratory tract problems by flight crew members.

Thus, flight crews must be acutely aware of their own limitations in assessing their proposed actions in a high-stress marginal approach and landing situation.

Why try to salvage a marginal approach?

Approaches are made under marginal conditions many times throughout the world every day with the crew and the airplane in complete control. The question is: why try to salvage a marginal approach? What to do when the approach itself becomes marginal? The question "to land or not to land" is posed to every airline pilot many times during his career. As the aircraft nears that altitude known appropriately as decision height or minimum descent altitude the pilot must judge the situation in order to decide whether to continue the approach to landing or to initiate a missed approach. This is the time when the pilot must assess the environment around him by evaluating the cues and information available. It is also when the pilot feels the stress of added pressure from his management and passengers to complete the landing at the scheduled destination and the ultimate responsibility for the safety of passengers, crew, and the aircraft. The goal of every flight is to get the airplane safely on the runway, at the scheduled destination if at all possible.

There are obvious pressures to press on and get the "wheels on the ground." For one, there is the commitment of the crew to maintain the on-time arrival so that the crew and the airline will look good in the schedule records and there is also the recognition of the unspoken needs of the passengers for a landing at the scheduled destination without the delays and inconveniences of landing at an alternate airport. Additionally, there is a general desire on the part of crew and traffic controllers alike to avoid having to carry out the procedures for a missed approach.

The pressure to continue also reflects an "ego" involvement associated with:

1. I can make it.
2. Why go around? I made it last time.
3. The aircraft ahead made it, why can't I?
4. I have a reputation to maintain.
5. Etc.

Add to this, the desire to get home and relax with the wife and children at the end of a long day or arrive in time for some planned activities, and scene is set for the high-stress marginal approach and landing accident or incident.

When pilots feel the added pressure to continue they should also consider the positive reasons for executing a missed approach. It is the pilot's responsibility to determine if the approach and landing are authorized under the existing weather minima in order to attempt a landing free of risk or danger. If the approach is being made in marginal conditions and it becomes necessary to execute a missed approach, it is because the pilot chose not to expose the passengers, the crew, the airplane, and himself to the hazard or danger that could possibly lead to loss or injury. In some situation, the pilot may determine that continuing the approach is unsafe or that he has doubts about the situation and environment. Marginal conditions sometimes cause the pilot to become rushed or to "get behind the airplane" as when ATC directs the pilot to turn the aircraft in close. A jet aircraft operating at low power in a high-drag configuration, descending at excessively high rates on final approach in marginal weather represents a prospective accident if the pilot is hurried or the approach is unstable. His job, the company's safety record, and the lives of people on the aircraft and on the ground, including his own, are at stake when the pilot chooses to continue an approach or make a landing that should not be made.

The pilot's judgment is the factor that determines whether or not to land. The pilot must determine if the conditions are such that a marginal approach exists, and to do this, he must have good prior flight preparation, knowledge of instrument presentations and sensitivity problems, and an understanding of meteorological illusions in order to form an opinion by weighing evidence and then comparing the situation with previous flight experiences.

Flight planning, especially during marginal weather, should begin before takeoff and be reaffirmed throughout the flight before the descent is made.

This provides the pilot with a detailed mental picture of the entire instrument approach procedure and runway environment before the approach is initiated. It enables him to fully concentrate on instrument interpretation and the actual execution of the approach. He should have a complete understanding of the missed approach procedures and be prepared to execute them. Not only is knowledge of the instrument presentations required, but changing instrument sensitivity while converging on a course often causes rather wild gyrations by pilots who have waited too long to properly establish their position on the glide slope and localizer. Trying to come back on course, when close-in, will aggravate the problem because of instrument sensitivity and the lower altitude and proximity to the runway. Thus, the precise execution of instrument approaches requires much practice. The pilot must also learn not to reject instrument references. To overcome the natural tendency to depend on spatial references, pilots must discipline themselves to trust instrument visual references and translate them into precise, safe instrument approaches.

As we have stated, weather often produces convincing illusions. Combinations of meteorological conditions and reflecting lights can produce deceptive illusions that often cause pilots to lock into false horizons, produce vertigo, and adversely influence the pilot's spatial orientation.

Many accidents and incidents have occurred after the pilot looking out called that he had the ground, the approach lights, the runway lights, or the runway "in sight," and the pilot flying the aircraft made the transition or was trying to make the transition from instruments to visual flight. Even after the runway environment has been sighted, the pilot is far from home free.

As the aircraft nears the decision height, the pilot has been accumulating the information in order to evaluate the situation and make an operational decision; to land or not to land?

At this time, descent and approach awareness are coordinated between the pilots, with the pilot not flying normally making necessary callouts. At decision height, the callous for "at minimums" is made. A wise pilot will have double-checked the decision height and computed and memorized the appropriate threshold altitude. The pilot flying will call out with a "missed approach" command if visual cues are insufficient. If things do not look right, going around and getting back up to where there is time and altitude to think things out is a very wise move, especially in marginal conditions. We must remember that it is the pilot's place to use good judgment; to assimilate, evaluate, and act upon the situation. Pilots must be reminded that executing a missed approach, when necessary, is a mark of good judgment!

Some technical aspects of approaches and landings under marginal conditions

Much has been said about margin over stall and coefficient of lift during all kinds of approaches: smooth air, high headwinds, crosswinds, turbulence, and so forth.

Equally well-covered are the various formulas for speed additives for these conditions along with wind shear and the techniques for coping with these phenomena. The reason speed additives increase controllability and margin over stall is obvious: the faster you go, the higher you are above stall speed and the greater the availability of aerodynamic forces for control. However, simply because one flies at 30 percent above the stall speed and adds 10 knots or so on top of that for gusts, he might not have all the margins he thinks he has while maneuvering to maintain a wings-level attitude plus trying to line up with the runway while in turbulence.

Figures 4 to 6 depict the degrading effect of spoiler deflection on the available margin over stall, the load factor to stall, and the change in pitch attitude or angle of attack to maintain constant lift during lateral-control applications for a typical jet transport in the landing configuration.

FIGURE 4 - DEGRADATION IN STALL SPEED MARGIN DUE TO LATERAL CONTROL INPUT

FIGURE 5 - LOAD FACTOR TO STALL DURING LATERAL CONTROL INPUT

[Figure: Graph showing pitch attitude or angle of attack (deg) vs control wheel deflection (percent). Note: pitch attitude = angle of attack if thrust adjusted to maintain constant flight path. Labels include "TYPICAL RANGE FOR MAXIMUM CROSSWIND OPERATION" and "MAXIMUM LANDING FLAP".]

FIGURE 6 - CHANGE IN PITCH ATTITUDE OR ANGLE OF ATTACK TO MAINTAIN CONSTANT LIFT AS A FUNCTION OF LATERAL CONTROL INPUT

The percent increase in the stall speed, pitch attitude, or angle of attack, to maintain constant lift, and the load factor to stall are presented as a function of control wheel deflection. One hundred percent control wheel deflection represents full lateral control application (full asymmetric ailerons and full spoiler deflection on one wing panel). It should be noted that these three factors all are additive when the control wheel is deflected so that for any given instantaneous wheel deflection, the resulting degradation in margin is not one or the other of these functions, but all three combined. For example, on a typical jet with maximum landing flaps, a 50 percent deflection would result in a 4 percent increase in stall speed, reduce the load factor to stall to about 1.58g from 1.7g, and require a change in angle of attack to maintain a constant lift of approximately 2 degrees. All this not only effectively lowers the margin to stall to 21 percent instead of the 30 percent plus 5 knots the pilot thought he had, but it also limits the angle to which a bank turn can be made before stalling and requires a commensurate increase in thrust and pitch to maintain the same lift.

We believe that the foregoing is not generally known by flight crews and may offer an explanation as to why such a large number of accidents and near accidents have occurred when the airplane landed short, landed hard, or hit the ground while attempting a missed approach under turbulent and/or otherwise marginal conditions.

Chapter 9

ENGINEERING / CONSTRUCTION

There are many facets to engineering where the atmosphere plays a role. If proper planning of anticipated atmospheric conditions can be performed, risk and liability are reduced. But, there are many occasions where we don't have the for-sight to accommodate such factors since they have very knowledge in climatology.

A case where there would be a factor; increased risk in roof damage would be in the shingle placement along the ridge of a roof. If the prevailing winds are from the west, then placement of the cut shingles should be placed from an easterly position and work your way to the west. The layering of the shingles would favor the wind/rain from the west and there would be less water damage or worn shingles in the future.

Digging a basement in a wet area: consider the water table prior to building a basement. If it's dry, you can't assume that you'll have an easy dig as water could "hydraulic," or work its way up through the floor. If a study was performed to determine average rainfall and drainage as well as type of soil, clay as opposed to a sandy sediment, this could have an effect to the overall performance of your dwelling and future problems.

There isn't enough room in the book to list all the problems that could develop.

Wind Issues

The forces imparted by wind onto buildings and other structures govern their design in many countries. Buildings are designed for strength, safety, and the comfort of the occupants in resisting both extreme wind storms (such as hurricanes/typhoons) and commonly occurring winds (such as gradient winds, downbursts, monsoonal storms and Chinook winds). On the other hand, environmental issues, such as the transportation of airborne pollutants, may come to the fore during times of calm or low winds.

Wind is the movement of air relative to the surface of the earth. It's caused by solar heating of the surface of the Earth which, in turn, causes pressure differences in the Earth's atmosphere and forces generated by the rotation of the Earth.

Some common types of wind events:

Monsoons are thermally produced seasonal winds that develop most strongly in and around the vast land mass of Asia.

Tropical Cyclones (also called Hurricanes and Typhoons) are intense cyclonic storms driven by the latent heat of oceans, generally developing at latitudes between 5 and 20 degrees.

Thunderstorms are caused by the upwards movement of warm, moist air which then cools rapidly and starts to sink, condensation then produces heavy precipitation (rain or hail) which falls, dragging cold air downward forming a downburst of wind.

Tornadoes are a funnel-shaped vortex of air producing the most powerful of all winds on Earth, and which typically occur in continental plains in countries such as USA, Russia, and South Africa.

Wind engineering uses a multifaceted approach to engineer a solution to wind issues and problems, and may involve various aspects of:

- The probability and statistics of the wind, structure, and the environment;
- meteorology
- fluid mechanics
- structural dynamics.

In spite of the tremendous amount of research in the field of wind engineering, wind loading design codes cannot be exhaustive in their coverage. The boundary layer wind tunnel has become the primary tool for use in studying wind issues and problems related to civil engineering applications. Wind tunnels are used to simulate the lowest layer of the atmosphere, known as the atmospheric boundary layer.

Wind Pressure on Walls and Roofs

Wind tunnel testing, to predict surface pressures acting on tall buildings and other complex structures like stadium roofs, for the purposes of cladding design, can result in a significant reduction in cladding costs. A rigid model of the subject building, typically with a length scale of 1:200 to 1:500, is fitted with a large number of pressure taps (typically 300 to 500) through which surface pressures are measured using a custom built high-speed pressure scanning system to reduce testing time. The simultaneous pressure measurements made possible with this system may also be used for the design of main structural systems.

CLP Power Wind/Wave Tunnel Facility, Hong Kong.

To ensure that a building is structurally safe and will perform efficiently once it is built, designers need to understand the nature and impact of the local wind environment. Wind engineering is the science that addresses this issue—it enables experts to model local wind behavior and then assess its effect on local terrain and buildings.

A wind engineering study is usually instigated when:

- Building falls outside the wind loading codes.
- More wind loading data is required.
- Planners require a pedestrian wind study as part of the Environmental Impact Assessment.

The earlier that wind-related issues are considered, the greater the opportunity to optimize design and to make savings. In many instances wind engineering will simply involve undertaking a desk study, but in complex cases it may require additional assessments: design reviews, wind tunnel testing and CFD modeling.

Desk studies

The wind consultant studies existing information and, in the light of their knowledge and experience, advises the client whether it is necessary to undertake a more detailed assessment.

Wind tunnel testing is the established and reliable method of assessing situations that require deeper investigation than a desk study. It allows designers to assess a number of key issues and optimize design before construction commences. This minimizes the risk of having to carry out costly and disruptive remedial work further down the line.

Tests are used to:

1. Determine structural loads, cladding loads and dynamic response
2. Predict whether the pedestrian environment will be comfortable and safe

3. Provide design data that can be used to optimize the size and location of inlets and exhausts for natural ventilation systems
4. Assess the performance of natural smoke ventilation or mechanical ventilation systems
5. Identify dispersion patterns of pollutants expelled from buildings
6. Test new and innovative systems

CFD

Computational fluid dynamics is becoming increasingly popular as a tool for predicting air flows and is particularly effective for assessing internal conditions such as ventilation air flow. However, it is not yet suitable for all investigations, particularly those involving the external wind environment where gusting occurs. In some instances, the best results are achieved by using CFD alongside wind tunnel testing.

The accuracy of the numerical (CFD) solution depends not only on the resolution of the spatial and temporal discretion schemes, but also on the treatment of various physical phenomena that can be represented only approximately in the mathematical model. Of these, the treatment of turbulence is often the most important in respect of the accuracy of the predictions. This applies particularly in wind engineering where the wide ranges of turbulent length and time scales, and the importance of gust conditions, means that great care has to be taken in the application of CFD and the interpretation of the results.

Structural and cladding design

In the case of structural and cladding design, the designer generally needs to know peak or gust loads. Fluctuating wind pressures are measured by fitting the model with 'pressure-taps'—brass tubes that end flush with the surface and are connected to pressure transducers. These transducers respond to wind pressures acting on the surface of the model.

Pedestrian environment

To assess pedestrian environment, 'time-averaged' and 'gust' wind speeds are measured using velocity probes mounted at pedestrian height. Flow visualization tests identify regions where high wind speeds are likely to occur at pedestrian level. Quantitative measurements are carried out at a number of locations where high wind speeds are anticipated, and where a comfortable environment is required (for example, parks, squares and pedestrian walkways). Measurements are analyzed with local meteorological wind data in order to predict the frequencies with which different wind speeds will occur. Safety and comfort criteria are used to assess whether the wind speeds are acceptable. If they are found to be unacceptable, wind engineers work with the designer to develop and test remedial measures.

Ventilation

Time-averaged measurements of surface pressures are carried out using techniques similar to those employed for measuring wind loads. Pressures are measured at locations coinciding with ventilation openings and exhausts. The pressures may be used as boundary conditions to predict air change rates; typically, the flow paths through a building (windows, cracks) are represented as a network of flow resistances. Pressures may also be used in CFD models to predict detailed velocity and temperature distributions within a room.

Pollutant Dispersion

The pollutant source is represented using tracer gases or smoke, which is released into the flow with the appropriate momentum and buoyancy characteristics. Measurements of tracer gas concentration are taken at key locations to assess dispersion patterns and pollutant concentrations. Acceptability will depend on a variety of issues including the toxicity of the pollutant emitted and odor. Any assessment is therefore carried out with reference to the impact of the pollutant, for which there may be guidelines or legislation.

Gas and particulate emissions from various industrial plants and vehicles can be transported by the wind to surrounding areas and buildings. These issues can be investigated on a micro scale in the wind tunnel to determine the dispersion of pollutants in areas of complex terrain (mountains, valleys, waterways, and buildings) as well as dispersion in street canyons. Both flow visualization and quantitative measurements (using fast flame ionization detectors) can be undertaken.

Dispersion of Energy
Courtesy: Hong Kong University of Science and Technology

Dispersion of vehicle exhaust.
Courtesy: Hong Kong University of Science and Technology

An example of Engineering details considering wind-flow:

(Europe) North Quay is a mixed office and hotel development in London Docklands. The building form is regular, but it is close to surrounding tall buildings, so an engineering firm was called in to measure the cladding loads, overall structural loads and the wind environment.

Testing

A static model of the tower and surrounding area was constructed in a test boundary layer wind tunnel, at a scale of 1:300, with more than 500 pressure taps. Ordinarily, a building of this shape would be standard in terms of wind loading; however the tall buildings upwind of North Quay have a significant effect on the wind loads, which means the code does not apply.

Results

Wind tunnel testing found that wind loads on the east façade of the tower corresponded with standards; however the wind loads on the west façade were much higher than the standards (over 100% higher in places). This was due to wake buffeting from the adjacent taller buildings.

Conclusions:

The design could be adjusted to reduce the overall loads by up to 20%, resulting in large potential savings in the cost of the structure. It also gave the client confidence in the design of the structure and the reliability of the cladding system. Wind tunnel testing therefore led to a safer and more serviceable design.

Testing:

A desk study was carried out to determine the heights of chimneys required to effectively disperse the pollutants.

Results indicated that nine stacks of 100 m in height were required. However this was not acceptable, both in terms of aesthetics and due to a height restriction. To resolve the potential conflict between the planners and environment agency, the client asked an engineering firm to undertake wind tunnel tests to determine the minimum stack height acceptable.

The wind tunnel tests will also be able to model and determine pollutant concentrations over the complex urban area in which the development is to be built.

Smoke analysis in an international airport:

The main issue was whether the proposed open-plan design for the arrival/departure area and the retail atrium space would be safe in the event of fire. It was proposed to use minimal compartmentation and structural fire protection, and to leave the HVAC system running in 100% replacement mode. Coupling a study of smoke and heat transport with predictions of human behavior, the engineering firm was able to provide the client with a complete integrated smoke movement and evacuation analysis. This was used to inform the design of the smoke extraction system.

The fire scenario used for the modeling studies was based on a luggage fire test, the engineering firm was able to demonstrate that the smoke layer in the departure/arrival area and retail atrium space would not cause a problem to the evacuating people, and that the structural elements would not be damaged by the heat. The integrated smoke movement and egress simulations confirmed the original design concept.

The wind study, engineering information was obtained in part from the following:

The CLP Power Wind/Wave Tunnel Facility : *Central Research Facilities* at the Hong Kong University of Science and Technology.

BRE is the UK's leading centre of expertise on buildings, construction, energy, environment, fire and risk. BRE provides research-based consultancy, testing and certification services to customer's worldwide.

Greg MacMaster
Delay Claims in Construction Cases: Common Pitfalls

Some of the most common disputes in construction cases relate to delay. However, delay claims tend to be some of the least understood and frequently confusing claims in the construction field. A clear understanding of the basic elements necessary to prove delay claims is invaluable in the processing of complex construction claims.

Much as it sounds, a delay claim on a construction project relates to a period of time for which the project has been extended or work has not been performed due to circumstances which were not anticipated when the parties entered into the construction contract. The most common causes of delay on a project include: differing site conditions; changes in requirements or design; weather; unavailability of labor, material or equipment; defective plans and specifications; and interference by the owner. Such delays will often force a contractor to extend its schedule to complete the work required under the contract, as well as to incur additional costs in the performance of said work. Generally, these costs may include: the costs of maintaining an idle workforce and equipment; unabsorbed office overhead; lost efficiencies; and general conditions. However, in order to receive an extension of time for project completion, or to recover additional costs, the contractor must meet a number of prerequisites.

A delay must be excusable in order to be the basis for an extension of time or additional compensation. Categories of excusable delay are often determined in the contract and typically involve matters beyond the control of the contractor. Examples of excusable delay include design errors and omissions, owner initiated changes, unanticipated weather, and acts of God. A non-excusable delay is a delay for which the contractor has assumed the risk under the contract. Oftentimes, even if a delay appears to be excusable, it will be the responsibility of the contractor if it was foreseeable; it could have been prevented but for the acts of the contractor, or it was caused by the negligence of the contractor.

Delays may be further classified into compensable and non-compensable delays. If a delay is compensable, the contractor is entitled to recover compensation for the costs of the delay in addition to time extensions to complete the project. Most contracts will include classes of delay which are compensable. The general rule, however, is that if the delay could have been avoided by due care of one of the parties, the party which did not exercise such care is responsible for the additional costs. The contractor may also be liable for the negligent acts of its subcontractors. If the negligent subcontractor is in the chain of privity with the contractor, the contractor cannot recover delay damages from the owner as those delays are the responsibility of the contractor. However, if the subcontractor has a direct contractual relationship with the owner of the project, the contractor most likely will be able to recover damages as it was not in a position to prevent the delay. Additionally, in order to recover damages, a contractor must show a link between the delay and the resultant damage. Simply stating that there was a delay is not sufficient without showing a nexus between the delay and the damages.

Even if it is able to meet the foregoing criteria, a contractor will not be entitled to recovery if there is a concurrent delay effecting completion of the project. A concurrent delay may be defined as a second, independent delay occurring during the same time period as the delay for which recovery is sought. If the party seeking increased compensation is ultimately responsible for the concurrent delay, he may not be able to recover any compensation for the initial delay. Some courts, will allow the aggrieved party to attempt to apportion the responsibility for delay, thusly allowing compensation to the contractor for the period of delay which was not its responsibility. See, e.g., Raymond Constructors v. United States, 411 F.2d 1227 (Ct. Cl. 1969). However, apportionment of delay is often difficult due to inadequate project documentation of the various delays. The best time in which a contractor can apportion delay is while the project is ongoing. Courts tend to find analyses made concurrent with the delays to be more reliable than after the fact analyses.

A contractor can do a number of things to make it easier for it to recover delay damages incurred on a project. The contractor should make sure the construction contract clearly defines items which the contractor will be able to recover. Additionally, each and every delay incurred should be well documented during the course of the project. Notice that the delay is impacting the contractor should be given to the party with which it is in privity. Finally, if there is any portion of delay for which the contractor is responsible, it should seek to apportion the overall delay between the items it is responsible for and those for which it has no responsibility.

Courtesy (*Scott A. Aftuck* of *Haese, LLC*)

Has El Nino Adversely Impacted Your Construction Project?

Winter rains have impacted almost every contractor. The abnormal weather conditions have delayed most projects and adversely impacted many contractors' productivity. Many contractors now ponder what rights they have to recover the additional costs that they have incurred as a result of the inclement weather. This article shall review which facts a contractor must establish to recover compensation for adverse weather conditions.

Weather Related Delay Claims-

In order to prevail on a delay claim, the party asserting the claim must prove that the delay was excusable, compensable and critical. If the contractor fails in establishing all three elements, it will not be able to recover additional compensation. However, if the contractor establishes that the delay was excusable and critical, it will be entitled to an extension of time.

Generally, most contracts do not excuse all weather related delays. Typically, the contract will only excuse rain delay days that were not foreseeable. Many government contracts now include an addendum that sets out the average rainfall for the anticipated construction period. For example, if during the scheduled period of construction, there were more than twenty days in which there was measurable rainfall, the contractor is only excused from performance on days with measurable rain after the twentieth day. In essence, these types of contract clauses require the contractor to assume the risk for all but unforeseeable weather-caused delays. Thus, it is necessary to first review the construction contract to determine how weather related delays are treated.

In most contracts, adverse weather delays are generally not compensable, but they will entitle a contractor to an extension of time. Nonetheless, if another preceding compensable delay extended the construction time into an adverse weather period, the weather related delays will be compensable. Similarly, if the owner requires the contractor to accelerate his performance to avoid adverse weather or to recapture the weather related delays then the contractor would be entitled to compensation. An example of additional compensation to recapture time lost to a rain delay is presently occurring at the San Francisco Airport. The airport commissioners are currently considering offering over nine million dollars in incentive, if the contractors can overcome the rain delays and complete the projects on time.

The final element of a weather related delay claim is establishing that it was a critical delay. Simply stated, the contractor must establish that the weather related delay adversely impacted the completion date for the project. Clearly, the completion date will not be impacted by abnormal rainfall when the only work that is occurring during the period is interior work. On the other hand, framing work would be impacted by abnormal rainfall. Thus, an analysis of how the work is impacted by the adverse weather is necessary.

Weather Related Disruption Claims—

While inclement weather may not delay the project, it could adversely impact the contractor's performance. For example, the adverse weather could result in an uneven labor force, inefficient labor force, out-of-sequence work, trade stacking, out-of-sequence material deliveries, and/or a loss of project momentum.

The determination of the right to compensation for a weather disruption claim is generally the same as that made for a weather related delay claim.

To prove a disruption claim, it will be necessary to identify the impacted activities. Specifically, the contractor should make contemporaneous records which establish which activities were impacted and the additional labor and costs associated with the disruption. If the contractor has failed to keep such contemporaneous records, there are other methods for proving the claim which include an engineered analysis, modified total cost, and/or a comparison analysis.

The disruption claim typically includes the additional direct and indirect costs for the impacted activity. The direct costs would include the increased labor, material and equipment costs. The indirect costs can include home office overhead and site overhead.

Documenting the Claim—-

Obviously, any determination of an excusable weather delay or disruption will require proof that the work was impacted by the weather. Climatological data can be obtained from reports prepared by the National Atmospheric and Oceanic Administration. The reports can be found in most public libraries or on the Internet. In addition to establishing the actual number of rain days and the amount of rainfall, it is also necessary to prove that the contractor's work was in fact impacted by the delay.

The contractor should also keep daily job logs that document the weather conditions. The daily log should identify what trades, if any, were impacted by the adverse weather. The log should also set forth how the weather disrupted the project's performance.

In closing, if you properly documented the adverse weather this winter, *El Nino* storms may not have been as disastrous as originally believed.

This article, ©1998, was written by William C. Last, Jr.

Chapter 10

AUTOMOTIVE ACCIDENTS AND IMPACTS FROM WEATHER DISASTERS

Probably the most common request you'll receive will somehow be associated with a car accident (depending on geographical location). Most of the calls I have received concern accidents from 12 months to 5 years ago and include anything winter can throw your way. Let's face it, winter storms are deceptive killers since most of the deaths that occur are indirectly related to the actual storm.

If there was ever an elective science class for law enforcement students and police officers, it should be a meteorology class. Since the weather plays a vital role in accidents, the first on the scene, after assessing for injuries, the officer should take an observation of surface and air conditions. Record air temperature, winds, type of precipitation, accumulation of precipitation on the ground—basically a complete weather report. (A similar report is recorded on notification of an aircraft mishap.) Condition of the roadway is also important. Go so far as to measure the amount of ice/packed snow on the road where the beginning stage of the accident started and ended. You can obtain a mobility weather station (hand-held)

Kestrel 4000 Pocket Weather Tracker from Ambient

from a number of on-line weather companies.

Re-occurring certification, bi-annually, on this subject would insure a higher degree of accuracy in reporting the weather on traffic citations.

If there's a disagreement on whether there was ice, check the scene of the accident for any skid marks. If it's been years, you may not have any physical evidence. Check around (conduct interviews) with local businesses owners, entrances where trucks may wait to enter—this may warm the ground where some ice could form after the truck leaves the spot. Work with other investigators and learn what they look for. This will only strengthen your sensitivity to notice minute details which could be a factor.

Edmond Locard's Principle of Exchange states that when any two objects come into contact, there is always transference of material from each object onto the other.

http://www.computing.surrey.ac.uk/ai/impress/ 06-19-2003

- People die in traffic accidents on icy, snow packed roads.

Transportation-related injuries can generate complex and expensive litigation, requiring extensive expert testimony, time-consuming discovery of defendants' paper and sophisticated navigation of the federal, state and local transportation laws and regulations that often control the result in a major transportation case. Check out this site to see a statistical listing of accidents.

http://www.car-accidents.org/stats-conditions.htm

These accidents can vary tremendously and it's up to you, the forensic meteorologist, to get a good idea as to the presumed cause of the accident as well as other factors that have been documented. The deeper you dig for answers, the better your chances of finding other factors that amplified the cause of the accident.

Here's an example:

It's a cold, winter day, snow bands have reduced visibility and some parts of the road had minor drifting (couple of inches). Winds were blustery in open fields and the temperature was quite cold, -22 degrees F. Speed limit for the road was posted 55 mph.
The accident report indicates that a driver spun out and hit another car and that there were injuries, although none severe—at that time. Two years pass and you're requested to gather data for a lawyer to determine how badly it was snowing and if there was ice on the roadway. This should sound familiar as it was discussed it in a previous chapter. At a glance, with pertinent data available (snowfall records, plow schedule, observation data from the closest station a few miles away) tend to learn towards a typical snowy day with blowing and drifting. Operator error, driving too fast for conditions or other decisions made by the officer that could show fault.
If you were to type a report of weather conditions at that time and preceding 6–12 hours, you would simply refer to the observational data sheet and compare with local plow records to substantiate the claim that there was snow on the ground. A one page expert witness report could easily be typed and sent to the requesting lawyer within a day or two. It's an easy couple hundred dollars, right? But the idea

that I'm trying to instill here is to look at all the possible ways the accident could occur (within the realm of meteorological explanation). Tree line, protected areas from wind, inclines declines, intersections or anything else that could cause a change in the apparent wind/weather condition. To see this means you'll have to go to the site and physically check yourself.

Intersections are another place where accidents occur and drivers have little knowledge about something we, as meteorologists, can see very well. Glazed ice! Here's how it forms. It's snowing and air temperatures are well below freezing, so there's no way for rain to form, right? Well, we have heat generated from car engines that melt the snow (to varying degrees). As traffic resumes, the melted snow then freezes. This process continues every 3 minutes from when the traffic light turns green to red. You can now see the process of freezing and melting for hours at a time. Glazed ice forming at intersecting roads makes a perfect combination for accidents to occur. When any two objects come into contact, there is always transference of material from each object onto the other (*Edmond Locard's Principle of Exchange*).

This is very evident when car tires carry amounts of salt and calcium chloride (ice melting). This exchange of material will have an effect on what is "traditionally pure water" on the ground. If the chemical composition of water changes, so does the freezing point.

Now on to sensors: Weather sensors are located 2 meters (almost 6 feet) from the earth's surface. It's a constant around the world. Every weather station that is recorded with the WMO, NOAA, NWS are all 2 meters above the ground. If the temperature is recorded to be 33F, 34F or 36F degrees, does that mean there was no ice on the ground? Cold air sinks (heavier) and if you did a study of air temperature verses ground temperature, you just might find a difference as high as 5–7 degrees (depending on mixing due to strength of winds). This could give credit as to the ice on the ground when temperatures were above freezing. Solar radiation and wind flow could offset this finding, so it's a good idea to know and understand the thermodynamic principles and formulas.

Understanding these formulas will give you better insight on how the atmosphere works and is a great way to find information from limited availability of data. Often used in depositions and courtroom testimony, the chart would show conclusive evidence with theoretical and sound reasoning relieving you of the "junk science" accusation. Determination of unreported meteorological quantities from plotted soundings (RH%, vapor pressure, mixing ratio, thickness, pressure and density altitude, LFC, LI, SSI, CCL, equilibrium level and many other values) can easily be definable and used in your interpretation of what elements may have been a factor. To learn more about the Thermodynamic Diagram, plotting and analyzing, write to *Cyclogenesis@charter.net* and request information on "How to Plot and Analyze a SKEW T log p Diagram." It will come in handy.

Tools and supplies needed on cases may vary, but it would help to have the following;

- Camera (digital preferably)
- Tape measure
- Distance measuring equipment (long rope, yardage indicator) to measure distance from trees or other obstacles at a distance
- Temperature sensors (pocket-size will suffice), used to measure 2 meters above the ground, at ground and below ground level.
- Pad of paper (graphic and lined)
- Topographic maps showing area to be surveyed
- Check the slope of the highway from centerline to edge, for drainage (or lack of).
- Make note of objects close to road that would block the sunlight (favorable for formation of black ice)
- Any other objects that seem out of place

As you might conclude, this list may be modified to suit your own personal investigation needs. But to get a good assessment of the location and variables, you'll need a good reference to work from. Supporting evidence showing lack of ice (black ice/packed snow/liquid) would be the presence of skid marks.

What follows is a detailed examination on speed/braking action and formulas.

SPEED CHART

$S = 5.5 \times \sqrt{(Cd \times Length)}$

[Chart: Speed (vertical axis, 0 to 70) vs. Length of marks (horizontal axis, 0' to 180'), with two curves for Cd = 0.8 and Cd = 0.7]

1. Determine the speed in mph after the vehicle has stopped sliding. This might be zero, but generally won't be zero. If the vehicle strikes something at the end of the skid, it is still moving and this residual velocity must be allowed for when using the chart.
2. Move out (horizontally) on a speed line until it hits one of the curves, the lower is for a 0.7 drag factor, the upper is for 0.8. This is the starting point for finding the speed drop associated with a given length of skid marks with a given residual speed after the mark ends.
3. Move out the length of the longest single skid mark, then move up until you hit the curve again to find the original speed.

EXAMPLES:

a. A vehicle leaves 40' of skid marks before coming to rest—the original speed is about 30 mph for both drag factors.
b. A vehicle leaves 60' of skid marks before striking another vehicle, the damage and post impact slide suggest that its speed was 30 mph when it struck the other vehicle.

1. Move out from the 30 mph line until you hit the curve at about 40'.
2. Add the length of the skid mark—60'—go out to the 100' mark and then up until you hit the curve again at about 46–48 mph. This is the original speed of the vehicle.

Marine accidents are different in nature in that most of the accidents that occur are from falling overboard or negligence on the part of the operator. However few the cases involving the weather, you still need access to information that could substantiate the claim of damage. Radar, lightning, satellite and local marine/surface observations are a few of the requests I have had to fulfill.

Time, Distance & Speeds Fundamental Relationships:

1. An easy one: Distance = Speed x Time -Example: 30 mph x 10 hrs = 300 miles
2. A useful one: 1 mph = 5280 ft/hr /3600 sec/hr = 1.47 ft/sec. So: 20 mph = 29.4 ft/sec.
3. **Acceleration:** The **distance** covered while accelerating for T secs.:

   D = Vo x T + 1/2 A x T xT where: Vo is the vehicle's speed when it starts accelerating (0 for a standing start) and A is the acceleration rate- for cars this is about 10 ft/sec/sec.

   Example: Vo = 0, T = 2 sec. D = 1/2 x 10 x 2 x 2 = 20 ft.
   The **time** required to cover a given distance from a standing start:

   T = sqrt (2 x D / A) - Many calculators will do square roots.

4. **Braking Distance:** The total distance required to stop the vehicle equals the distance covered while the driver is detecting and reacting to the situation plus the distance covered while braking. So D = Vo x T + the braking distance:

Db = S x S /(30 x Cd) where:

S is the speed in mph and Cd is the drag or braking coefficient, 0.7–0.8.

Braking time: Take the speed in ft/sec and divide it by the deceleration rate- about 0.7 -0.8 x 32.2 = 22- 26 ft/sec/sec. Add the reaction time.

5. Speed from skid marks:

If the vehicle is stopped at the end of the skid marks, then the original speed is : S = 5.5 x sqrt (Cd x Length of skid marks) where the Cd is as defined above and the speed is in mph. If the vehicle is still moving when the skid marks end, this formula does not yield the original speed. (*see the "Combined Speed" sheet*)

6. Speed from damage:

It is possible to compute speed changes from crush damage but the calculations are complicated and generally require a computer along with damage measurements and knowledge of the structural properties of the vehicles involved.

The Combined Speed Formula

$$S^2 = S_1^2 + S_2^2 + S_3^2$$

- With a multiple phase accident where a vehicle may have undergone several speed changing events, it is necessary to use the combined speed formula (above) to estimate the original speed of the vehicle.
- If for example, a vehicle loses 30 mph in sliding before impact, 20 mph in the collision and 10 mph in sliding to a stop after the impact, the original speed of the vehicle is then:

S x S = 30 x 30 + 20 x 20 +10 x 10 = 900 + 400 + 100
= 1400, so that S = 37 mph

Notice that we cannot simply add the speed decrements together, they must be combined using the formula. This frequently leads to lower speed estimates than might be anticipated when looking at the various speed numbers before they are combined. The reason is that speed energy—what's really being computed in the partial estimates—varies with the square of the speed. Thus 20 mph is four times the (kinetic) energy of 10 mph (20 x 20 =400, 10 x 10 =100), not twice as much. The speed from skid marks formula - S = 5.5 x sqrt (cd x distance)—is a computational aid which assumes a zero final velocity, that is, that the vehicle has stopped at the end of the skid marks; if the computed value is to be taken as the original speed of the vehicle. It does not yield speed drops simpliciter. Thus, 20 feet of locked wheel skid marks represents about 20 mph if the vehicle is stopped at the end of the marks. Suppose however that the vehicle is doing 60 mph when it leaves 20 feet of skid marks. Its speed at this point is: S = sqrt (3600 - 400) = sqrt 3200 = 56.6 mph.

Weather Impacts in Canada

(Environmental Adaptation Research Group, Atmospheric Environment Services Downsview, Ontario, Canada)

Canada is subject to a variety of hydro-meteorological hazards and disasters. An inventory is being assembled and is summarized in *figure 1*. This chart shows the number of identified events by hazard for which cost estimates could and could not be made. The total of known costs is identified in 1995 dollars at the end of each bar. Tornadoes are not included, since there have been over 2,300 known Canadian tornadoes during the period 1961-1996. I would also note that a precise criteria was not used to include or exclude particular events but rather events were included if the information source suggested a "significant" impact of some sort (meteorological, social or economic).

Figure 1 - No. of Identified Events

Hazard	Cost
Heat	No estimate
Hurricanes	Hazel -
Forest Fires	$0.5 billion
Floods	$1.6 billion
Cold	No estimate
Droughts	$7.9 billion
Storms	$1.1 billion

Number of Events (0 – 120)

■ Number events with $ estimate ■ Number events without $ estimate

Droughts, by far, are the most costly hazard, although they rank fourth in terms of frequency. It is worth noting that for hurricanes, only economic costs associated with Hazel (1954) have been included. An historical survey of Canadian disasters shows that 44% of them are weather or climate related, that almost 1/3 of all disaster has occurred at sea, and that 80% of those were weather-related.

Natural disasters are the extreme of natural hazards and occur when social vulnerability is triggered by an extreme event. The costs Canadians incur from hazards are a function of our adaptive decisions. Unsafe conditions result from a number of social forces which are rooted in limited access to power, economic resources and the nature of political and economic decisions.

These costs of natural hazards can be broken down into social costs and economic costs.

Social Costs

In terms of social costs the following statistics give some indication of the overall magnitude:

Transportation

Decreasing trend of weather-related aircraft accidents from approximately 60/year in 1985 to less than 20/year since 1992.

Although there have been few fatalities, there has been a significant number of weather-related railway accidents averaging between 20–40 per yea; however, 100–120 occurred in 1990 and 1991.

Historically, many of Canada's worst disasters have been ocean-based with typically 300–400 weather-related marine accidents each year (note 1990 worst year in the past 10 years with slightly less than 500 weather related marine incidents).

Major road accidents result mainly from wet conditions, followed by ice, snow, slush and mud. In 1992 weather-related road accidents resulted in 298 fatalities, over 23,000 personal injuries and over 72,500 property damage incidents.

Deaths

Canadians occasionally die as a result of atmospheric hazards. Most deaths occur as a result of cold. In the past decade, however, the number of deaths from cold has shown a gradual decrease, while those resulting from other atmospheric causes have remained fairly constant (note: some concern as to how attribution of cause has been assigned).

Economic Costs

There are two fundamental economic costs associated with natural hazards—adaptive costs and impacts, response and recovery costs. Adaptive costs are those associated with protection, reduction of vulnerability or risk, education and research. These costs are difficult to estimate and little research has been devoted to increasing our understanding of these costs. One preliminary estimation of Canadian adaptation costs suggests that $13.7 billion is spent annually; however, this is likely underestimated by some unknown amount.

Impacts, response and recovery costs are those incurred when protection fails or no adaptive response is taken. Some examples for Canada include:

Forest Fires

These can have a direct impact due to the loss of natural resource, though it is unclear how to account for these losses as fires are now considered an essential part of the natural ecological cycle. The annual area burnt suggests an upward trend with 1995 being by far the worst year with over 7 million hectares burned, followed by 1989. This statistic is related to climate, but is also related to decisions made regarding fire management. All provinces incur costs related to fire management. Over the last decade, Ontario spent over $800 million, more than any other province.

Annual fire management costs are shown on *figure 2* with costs peaking in 1995 at over $450 million.

Figure 2 - Fire Management Costs

Hydro Companies

Weather-related costs are highly variable as indicated by those for Ontario Hydro which has annual weather-related costs ranging from zero to $3 million and averaging $1.4 million/year.

The most costly hazard was "wet snow + high winds" the total cost which occurred due to one event at Vegreville and Lloydminster, Alberta. This two-day storm destroyed 108 steel transmission towers, 300 wood transmission structures, and more than 3000 wood distribution poles. In addition more than 400 km of conductor had to be replaced.

Tornadoes (numbering 8) come second in terms of costs, though they were all from one utility company.

Federal Payments

Floods have cost the most, by far, with audited totals over $300 million and federal payments of $150 million (does not include the 1996 Saguenay flood).

Storms and fire rank second and third with about 11% each.

Provincial Costs

Provinces incur costs due to crop damage from hail, flood, drought and a variety of other hazards.

The largest annual expenditure occurred in Saskatchewan, with payments totally over $500 million (1988 and 1989).

Drought is the major hazard resulting in crop insurance payments on the prairies (e.g., Manitoba paid out $400 million from 1966–1994 in response to drought-related crop losses, followed by excess moisture, hail, heat and frost).

Insured Costs

Figure 3 shows the costs of multiple major payments from 1983–1994 (does not include the costs of events less than about $4 million, and therefore the true costs are much greater than those shown in this figure).

Hail has caused the most damage (over $450 million), followed by tornadoes, flood, storms and wind.

There appear to have been 9 events in 1995 which include significant damage from flood, hail, thunderstorms, wind and Hurricane Hortense.

The largest single disaster in Canada was the Calgary hailstorm of 1991, which cost insurance companies around $400 million. The insured costs of the Saguenay floods are currently estimated at $350–400 million.

Figure 3 - Weather Related Insurance Costs
Major Multiple Payouts

1996 - at least $645 million

Canadian Weather-Related Disasters in 1996

In 1996, Canadians suffered through some of the most extreme and destructive weather ever to have hit the country. For most of the year, the weather either froze, buried, soaked, buffeted or frightened Canadians. No part of the nation seemed to escape the wrath of the weather in 1996.

Total property damage will likely exceed $2.5 billion when the final figures are tallied. Indirect costs and losses from revenue shortfalls, canceled events, missed opportunities, and slowed business will probably be in excess of a $3 billion hit to the Canadian economy. Remarkably though, the number of personal injuries and fatalities linked to weather incidents could have been much higher. Unofficial numbers point to fewer than 25 weather-related deaths (excluding deaths from road accidents and hypothermia)—10 from the storm in the Saguenay and six from lightning in separate incidents. The top weather stories of 1996 are as follows.

The Saguenay Flood

By far the worst catastrophe of the year, and Canada's first billion dollar plus natural disaster, was the flooding and mud slides in Quebec's Saguenay River valley in mid-July. The storm produced the largest ever overland deluge in Canada this century—an amount equivalent to a two-month flow over Niagara Falls—triggering a surge of water, rocks, trees and mud that killed ten people and forced 12,000 residents to flee their homes. It was the deadliest flood since Hurricane Hazel in Toronto in 1954.

The scale of the tragedy was staggering. Many of the region's roads and bridges and delivery systems for power and water simply disappeared. To the insurance industry it was Canada's worst-ever weather disaster in economic losses. By including insured and uninsured losses and indirect costs to the economy, total losses are sure to exceed $2 billion.

The Pacific Storm of 1996

During the period December 22 to January 3, 1997 a series of brutal winter storms blasted Vancouver Island, the Lower B.C. mainland and the Fraser Valley with up to 85 cm of snow paralyzing infrastructure and commerce. The accumulation of snow for this series of storms was unprecedented in historical records of Vancouver Island and the lower south coast of British Columbia. The recorded 64.5 cm snow that fell in a 24 hour period, at the Victoria airport, gave Victoria the distinction of having the 3rd highest snowfall, in a major city, in Canadian history. During the same period, downtown Victoria received 85 cm.

The environmental impact was severe in that massive releases of untreated of partly treated effluent were released into rivers and the ocean.

Although estimates of the economic losses range in the area of $200 million, the true impact of the storm will probably never be known. The following list suggests the overall impact

- over $200 million in total losses estimated
- largest search and rescue team in the history of B.C. dispatched
- 170 motorists stranded and later rescued
- 1700 people received emergency food and shelter
- Trans Canada highway closed for two days - stranding thousands
- secondary arteries closed for up to 10 days
- police and fire departments unable to respond to emergency calls
- 150,000 homes without power
- roofs and buildings collapsed under the weight of heavy snows
- over 500 avalanches reported
- 65 reports of pollution spills
- hundreds of flights canceled
- ski resorts closed
- theaters and sporting events canceled
- food shortage

High Energy Costs

In much of Canada, 1996 featured one of the longest and most vicious winters in recent memory. Three straight weeks of frigid weather gripped almost the entire country in January making it colder in most cities in western and central Canada than it was at the North Pole. To keep up with the cold, utility companies pumped out power in record amounts from British Columbia to New Brunswick. Canadians paid an additional $500 million to keep their dwellings as comfortable as in the previous winter.

Costly Prairie Hailstorms

In July, hailstones the size of fists bombarded Winnipeg and Calgary, racking up close to $300 million in property losses. In Manitoba, more than half the losses were for auto damage, making it the worst single disaster claim against the Manitoba Public Insurance Corporation in its 25-year history. At least a third of the cars damaged had to be written off. In Calgary, hail and flooding rains knocked out the city's 911 service and swept away cars.

Wet and Cold Weather Reducing Crop Yields

Unfortunately for western farmers, prospects in early September for one of the most bountiful grain crops in Canadian history didn't exactly materialize. Fall temperatures across the west were much below normal (the prairies had their second coldest fall in half a century) and precipitation was much above normal (the 7th wettest fall in half a century). Cool wet weather during the harvest of western red spring wheat led to a severe drop in its grade distribution, denying farmers an additional $180 million.

In southern Ontario, winter wheat production was severely affected by the wet cool weather throughout the growing season. Record rainfalls resulted in the worst outbreak of blight fungus ever seen in Ontario. According to Agriculture Canada, the excessive moisture and disease not only reduced yields, but it also reduced the quality of most of the crop to feed, since affected grain cannot be used for human consumption. The loss was estimated to be about $90 million.

Deep Winter Snows

So much snow fell early in the winter that before 1996 even started, many cities in western and central Canada had all but exhausted their snow removal budgets. Hardest hit was the central Ontario snowbelt from Barrie to Sault Ste. Marie, where on several occasions, cars disappeared in snow drifts, service centers became refugee camps, roofs collapsed and schools closed.

Insurance claims paid in the first three months of 1996 were 11% higher than in the previous year when the weather was much less severe. Total insured losses owing to the weather estimated at $165 million.

Slow Spring Affecting Retail Sales

For most of Canada, the winter season gave way to the monsoon season. Unrelenting rains and dreary weather plagued the country from April to June. Garden centers and golf courses were virtually empty during the spring. Retailers blamed the persistent cool and rainy weather for a 30% drop in the sales of weather sensitive goods and services, such as pools, air conditioners and warm season apparel. Sales of general merchandise in April and June were down by $100 million over the previous year's sales.

Flash Flooding in Ottawa and Montreal

The third major storm in less than two weeks, and the worst on record, hit Ottawa-Hull in early August with a deluge of 100 to 150 mm of rain in 90 minutes. Total insured property damage exceeded $20 million, not including the costs of repairs to damaged sewers and roads. Between November 7 and 9, thirty hours of steady rains drenched parts of Montreal and southwestern Quebec. The rains washed out sections of highways, collapsed bridges, derailed trains and undermined road and rail beds. Damage estimates put the event at over $60 million.

Severe Thunderstorms and Tornadoes

The snow had hardly melted in southern Ontario when the season's first tornadoes tore through regions east of Lake Huron in April. The twisters injured two people and caused property losses, much of it uninsured, approaching $8 million.

Severe thunderstorms on July 4 spawned at least eight tornadoes in Saskatchewan. Winds of 140 km/h and hail the size of golf balls produced $15 million in property damage. Two weeks later seven tornadoes touched down in Alberta, trashing trailers and flattening granaries to the tune of $10 million. Near Stoney Plain, more than

100 mm of rain fell in severe thunderstorms backing up sewers and flooding basements for another $10 million in losses. Tornado-related damage in Canada easily exceeded $50 million.

Spring Flooding

Significant flooding occurred in several communities across Canada during much of the spring and early summer. The Okanagan experienced its worst flooding in six years, The Red River inundated farm fields, roads and major highways leading authorities to declare a provincial flood disaster for the first time since 1979. In Winnipeg, the costs of filling 336,000 sandbags and protecting pumping stations alone cost $1.2 million. In Timmins, Ontario the Mattagami River overflowed its banks in the worst flooding across Canada and losses are still being tallied, with final figures are expected to range between $20 and $50 million.

Hurricanes and Weather Bombs

Four hurricane-force storms struck Eastern Canada in 1996: Bertha, Edouard, Fran and Hortense. It was the second consecutive season with above average hurricane formation in the North Atlantic. In 1996 there were 13 named storms of which 9 were hurricanes, including six intense ones, compared to a normal of 9 storms, 6 hurricanes and 2 intense ones.

Hortense, which swept east of Halifax and traversed western Newfoundland on September 14, was the first hurricane to achieve landfall in Canada in 21 years. Winds topped 161 km/h on cape Breton Island, felling trees, lifting roofs and blowing out windows. Total property losses approached $5 million.

Described as the worst storm since Typhoon Freda in 1962, a "weather bomb" struck Vancouver Island on October 17 causing massive power outages while felling trees, setting adrift 50 pleasure boats and ripping apart docks. A "weather bomb" is a storm which intensifies very quickly and moves faster than a hurricane. This storm packed winds as strong as 161 km/h and produced waves as high as 30 meters.

Conclusions

Are the number of weather-related disasters in Canada increasing? According to the Insurance Bureau of Canada, ever since the Edmonton Tornado of 1987, the number of multi-million dollar losses from weather disasters has been on the rise in Canada (note: before 1987 there was no single natural disaster with damages exceeding $1 billion anywhere in the world, let alone in Canada).

While the outburst of extreme weather was interesting, climatologists were not generally surprised by them. Climatologist are, however, becoming increasingly concerned that the volatile weather in 1996 might be a dry run of extreme conditions we might expect from a warming climate.

With this in mind, planners and decision-makers should note that natural hazards and disasters are expensive but not inevitable. With appropriate planning to reduce vulnerability, their social and economic impacts on Canadians can be reduced.

Retail Sales

Why do retailers continue to blame the weather for poor retail sales?

As retailers post annual results, the UK's weather gets the blame once again. Retailers cannot continue blaming the weather for poor business performance they must begin managing weather related risks. The correct weather forecasting products and services could soon make the 'wrong weather-poor performance' excuse invalid.

The impact of weather on retailers: Research by Datamonitor states that UK grocery and clothing retailers could be missing out on as much as £4.5 billion per year by failing to include weather forecast information in their short and long term business planning decisions. There are numerous examples of unpredicted weather hitting bottom line retail results.

House of Fraser was reported in Retail Week, June 2003, as stating: 'According to analysts, the Iraq conflict, followed by poor weather in May contributed to a tough trading climate on the high street.'

A sales and trading update from Thorntons, specialty retailer and manufacturer of high quality chocolate, stated: 'The exceptionally hot weather significantly reduced sales in the week leading up to Easter, although all egg stocks have been sold at discounted prices.' (Trading Statement by Thorntons, May 2003).

In April 2003, BBC News reported: 'Scotland's retail sector has dramatically outperformed the rest of the UK. Unseasonably warm weather in Scotland helped to lure Scots out for retail therapy.'

In October 2002, BBC News also reported: 'Next and Ted Baker have reported a surprise slowdown in sales, blaming unusually warm autumn weather. Both companies stated the Indian summer sunshine had deterred shoppers from buying their autumn and winter ranges.'

However, the UK's weather is not always a problem. For some retailers, the weather has provided significant gains. Burberry for instance, reported that the cold and wet weather over Christmas 2002 helped to drive trade in hats, scarves and coats (BBC News, Jan 2003).

Fashion retailer H&M, also benefited. In November 2002, BBC News reported: 'Sales jumped by 26% in October compared with last year. Like most fashion stores, H&M suffered due to unusually warm weather in September, which meant that people were not buying autumn clothes. However, a cold weather spell in October had helped to boost demand.'

Classic examples of UK weather impacting product sales include ice cream, chocolate, de-icer and umbrellas, which are clearly weather sensitive. Other examples are less obvious. For instance, each summer, UK supermarkets sell more toilet rolls than at any other time of year—around 20,000 more packs each month. The reason is that people suffer from hay fever during the summer and use significantly more tissues.

Managing weather related risks: Most retailers see improving supply chain and business-response lead times as ongoing goals; the next step is to make quick process changes in response to weather conditions.

More retailers are considering insurance to cover weather risks. It is possible to trade in weather derivatives to provide financial protection from unpredicted losses due to weather. However, purchasing weather related insurance is a new concept and few retailers have the knowledge, skill and experience to select the correct options. Some are following the direction taken by some of the worlds' leading retailers and adding weather forecasting and consumer demand models to the business decisions process.

Accurate weather information could improve the timing of key phases such as the introduction of high-summer fashion clothing. Allocations and store replenishment could benefit from moving forward the production, intake and allocations of summer clothing. Special promotions and advertising could be closely linked to weather forecast information and predictions.

What are the benefits for retailers that use weather forecasting? The benefits depend on how vulnerable products and services are to the weather, and whether forecasting information can be applied rapidly enough to allow retailers to make material changes to their business. However, it would be reasonable to expect the benefits to include increased sales, increased product availability, reduced markdowns and increased margins.

Retail Week recently stated that Safeway acted on accurate forecasts for hot weather in the run up to Easter 2003 and successfully modified their in-store promotions to deliver an extra £2million of sales.

What are the options? There are a number of organizations offering services for short, medium and long-term weather forecasts. They analyze the relationship between product sales and weather elements, and predict future consumer demand. These solution providers vary from institutions such as the

Meteorological Office, through companies such as Planalytics in the USA, to small UK-based partnerships such as Weather Commerce and Smart Research.

Making the most of weather forecasting techniques. Retailers must first understand their business requirements to use weather forecasting techniques to improve bottom line results. This means recognizing how their inability to predict the weather has impacted the business. It also demands an understanding of the processes and systems that require improvement.

It also needs a detailed knowledge of weather dependant product ranges and how they will be affected. Timing is also critical. Could retailers make business improvements based on weather forecasts up to 10 days, 30, 90 or over 90 days? There will be an optimum window and the trick is to identify it.

Weather related forecasts can improve business planning, supply chain and merchandising decisions. The potential benefits of managing weather related risks could offer significant return on investment. In the future, shareholders may begin to lose patience with retailers who continue to cite the vagaries of the weather as an excuse for poor performance.

Retail Sales (Courtesy: Paul Mason Consulting. RFID)

Chapter 11

SAMPLE CASES FROM THE FIELD OF FORENSIC & METEOROLOGY

Some of the cases below are from various sources and does not mean that we endorse them. We appreciate all material that has been submitted and they have all been credited where appropriate.

The cases below are real-life examples of how a forensic meteorologist can help make a difference in the legal realm.

A 24-karat weather verification

A jewelry store in the southwestern part of the United States ran a contest that stipulated if the temperature reached a certain record level on the day of the contest, that every customer who purchased from the store that day would receive free jewelry. The record temperature on the day of the contest was 115 degrees. According to the jewelry store, the temperature that afternoon hit 116 degrees, which meant that every single customer was given free jewelry, costing the store some $100,000!!

Prior to the contest, the store had taken out an insurance policy to cover their potential loss, should the record-breaking event occur. And according to the store, it did. They even turned in an "official looking" document to their insurance company, claiming they had a loss which they wanted reimbursement. The claims adjuster for the insurance company contacted a forensic meteorologist and requested a verification of the temperature for that day. The report provided to the insurance company by the forensic meteorologist showed a temperature of only 112 degrees. The reporting station that the jewelry store used in their claim to the insurance company was from a location which is not monitored or controlled by The National Oceanic and Atmospheric

Administration. Since there is no quality control at this site, it is not considered an official site for weather records. Based on the report, the claim was denied saving the insurance company $100,000.

Was it an event-in-progress?

Plaintiff, a woman in her 30s, filed suit against a hospital in southeastern New York, after slipping and falling due to a puddle of water at the entranceway to the building. The woman was entering the building with her boyfriend, who was there undergoing blood tests. Her boyfriend testified that he noticed the puddle at 9:20 am as they were entering the building. They both claimed that the accident occurred as they were leaving, and that it was not raining or snowing at the time of the slip.

Defendant Hospital contended that the accident occurred as plaintiff was entering the building, and that a snowstorm was still in progress at the time. At trial, expert testimony from a forensic meteorologist, confirmed the Hospital's stance that snow was indeed still falling at 9:20 am, and was part of a storm which had started several hours earlier. The storm came to an end shortly before 10:00 am, which was determined to be after the time that the slip took place.

Verdict came back in favor of defense, based on the forensic meteorologists' testimony that at the time of the slip, snow was falling, and thus, it was an event in progress.

When did the hole become visible?

Plaintiff, a woman in her 40s, filed suit against a horse farm located in the Catskill Mountains of New York. She sustained severe injuries after the horse she was riding on, stepped into a hole located on a riding trail inside the expansive farm, and fell over, causing her to be thrown to the ground. The trail is maintained by the farm, and is open year-round. The incident took place in the month of December.

Defendant Horse Farm contended that they had no knowledge of the hole's existence, as it was filled and covered by snow and ice from winter storm events which had occurred in the two week period leading up to the day of the incident.

At trial, the forensic meteorologist testified that it had indeed snowed on two different occasions in the 10 day period leading up to the day of occurrence. However, temperatures and degree of sunshine during subsequent days after the last snow event would have been sufficient to cause the amount of snow that had accumulated, to melt away completely. And that this would have taken place at least 60 hours prior to the date and time of the incident.

Verdict came back in favor of plaintiff, based on the testimony that more than two days time would have passed in between the time that the hole became visible after the snow melted away, and the time that the accident occurred.

Real-life examples provided by Compu-Weather.

The cockleburs on a ski-mask, that convicted a rapist

One midnight in midsummer in a suburb of Chicago a woman parked her car and walked toward her apartment building. Suddenly a man wearing a ski mask leaped from the shrubbery, attacked her, and then disappeared. The police began to suspect one man in the building, and with a warrant they searched his apartment and found a ski mask, which he claimed he had not used since the previous winter. The victim identified the man in a voice lineup, but this was not enough for a conviction. There were two cockleburs stuck to the ski mask and the detectives sent them to forensic entomologist Bernhard Greenberg for examination. Within the cockleburs were live weevil larvae. Examination of the cockleburs found on the crime scene proved to be of the same species as the weevil found on the ski mask. The species was identified as *Rhodobaenus 13-punctatus* Illiger, and is also known as the billbug. This species has a 1-year life cycle, and the larvae pupate in the cocklebur and emerge in the latter part of the summer, and then hibernate. Larva do not over winter, and they would not survive the winter within a desiccated cocklebur in a heated apartment. The suspect was then caught in a lie. The court trusted this evidence, and the rapist was convicted.

The chigger bites that convicted a man for murder

In 1982, deputies of the Ventura County Sheriff office noticed that a murder suspect had chigger bites similar to the ones investigators at the crime scene had on their waste-lines, ankles and behind the knees. The entomologist Jim Webb was contacted, and by analyzing the bites, Webb connected the suspect to the crime scene where the naked body of a 24-year-old woman was found on August 5, 1982. She had been strangled with her own blouse.

They did several tests at different places, but the only place they found chiggers in was a narrow strip near a eucalyptus tree under which the woman had been found. This meant that the suspect had to be at the crime scene at some point, which did not correlate with his testimony. The suspect claimed to have seen the woman the last time at a bar. The suspect was convicted for first-degree murder and sentenced to life without parole.

The partly submerged woman in the sandpit

The dead body of a woman was found in an old sandpit in the city of Helsinki, Finland on 27 June 1964. The sandpit was filled with water and the body was partly submerged and covered by bits of board. Presumably, it had earlier been completely under water. It was largely decayed, but still retaining its original shape. The soft parts of the face had decayed away. Adipocere formation had occurred. At autopsy on 29 June 1964, some fly larvae and puparia were detected in one hand. The larvae were dead and dry by the time they reached the Zoological Museum for examination. In spite of this it was possible to see that three of the larvae belonged to the genus *Muscina* and one to the genus *Fannia*. No flies or fly parasites had emerged from the puparia. No blowfly larvae or water insects were found.

Conclusions based on entomological findings:

The occurrence of larvae and unhatched puparia in the body indicate that it had been accessible to flies for at least one week, because in the prevailing cold micro-climatological conditions the development of *Muscina* fly egg to puparium takes about that time or a bit longer. The absence of blowfly larvae, which attack bodies in their initial phase of decay, indicated that the body had not been accessible to blowflies at that stage. Meanwhile, decay had advanced so far that the corpse was no longer attractive to blowflies, but was attractive to *Fannia* and *Muscina*. Hence it was concluded that the body had been completely submerged for a comparatively long time.

Validity of conclusions:

Police investigations showed that the woman in question had been murdered in the middle of July 1963 and had been hidden by the murderer in the sand-pit. Therefore the conclusion based on the absence of blowflies, although valid, had only a low degree of accuracy. The conclusion based on the occurrence of larvae or puparia of the genera *Fannia* and *Muscina* was indicative of the time during which parts of the dead body had been accessible to fly oviposition, but was of no significance for the determination of the time of death. Obviously, the hand had first emerged from the water and the flies had then oviposited on it.

The body in the bed

The body of a woman was found in her bed in a flat in the centre of Helsinki, Finland on 1 September 1965. Death had obviously occurred on 10 August, since newspapers had not been removed since that day. This was a case of suicide with sedatives. The cadaver was moderately decayed and greenish. The skin surface was loose and the viscera decayed. Blowfly larvae emerged from the orifices of the body. Autopsy was performed after regiration for one day and the fly larvae were collected on this occasion. The larvae were in bad condition at the start of the rearing, except for two small specimens. Rearing at room temperature yielded two small imagines of *Calliphora vicina* on 27 September.

Conclusions based on the entomological findings:

The woman in question had been dead for more than 7–8 days, because fly oviposition occurs on about the second day after death, and development to mature migrating larvae takes about 5–6 days.

The cadaver had been in shadow, because the scotophilic *C. vicina* had oviposited on it, but *Lucilia* species had not, although they were still active.

Validity of conclusions: Both entomological conclusions were validated by the known facts. The conclusion that more than 7–8 days had elapsed since death is true as such, but of little value, because death had in fact occurred 20 days before discovery of the cadaver. A better result would obviously have been obtained if living full grown larvae had been taken for rearing, or if puparia or adult flies had been collected by policemen in the flat of the woman in question.

The two murdered hitchhiking girls

On 21 August 1971, at 1600, the corpses of two murdered girls who had been hitchhiking were discovered in a sandpit near the town of Hyvinkaa, in southern Finland. The corpses were partially covered by a polyethylene sheet. A cluster of fly eggs was collected from the hair of one of the girls; a fly larvae between 4.5 and 5 mm long was also present in one eye. Four days later, examination of the refrigerated bodies revealed four larvae 5–6 mm long in the eyes of the same girl, and five larvae 2.5–3.5 mm long in the eyes of the other girl.

An attempt was made to rear all the eggs and the larvae to adult flies. The development of the eggs into larvae 4.5–5 mm long (i.e., to the length of the fly larvae observed on 21 August) required one and a half days. Further rearing was only partially successful and a single adult fly of the species *Calliphora vicina* was obtained. Flies of the same species were also obtained from a liver growing-medium placed in a glass container on 28 August 1973 in the place where the dead girls had been found.

Since it had taken one and a half days to obtain a larva 4.5–5 mm long experimentally, it was concluded that the bodies had been in the locality where they were found from 19 August, namely for about two days after the time of death. The suspected murderer, however, had an alibi for 19 August and the following days as well. During the trial the question was raised whether it was possible that the dead girls could have been in the place where they were found on 14 August, as suggested by police investigation. The answer to this question was that, considering the daily temperature from 14 to 19 August (well above 16 degrees Celsius during each day), a large number of big fly larvae should have been found in the corpses. Since this was not the case, one had to draw the conclusion that either the corpses of the girls were not at that place on 14 August or they were completely covered by the polyethylene sheet. From the photograph taken by police immediately after the discovery of the corpses, it seemed possible that the polyethylene sheet at first had covered the girls completely, but later had been partially removed by the wind. The subsequent question was whether fly oviposition could have occurred through possible holes in the polyethylene sheet. The sheet was immediately inspected, but no holes were detected. The results of the police investigation, substantiated by the entomological observations, led to the conviction of the suspected murderer.

Cases from the archives of: *Morten Starkeby* on Entomology

Chapter 12

RESOURCES, WEBSITES, COMPUTER PROGRAMS & SPREADSHEETS

Why re-invent the wheel, right? Since some of us may not have access to data, programs or reference link like others who have been in the business over the years, I decided to make my list and then if you find a site, program or reference that you feel is beneficial to the forensic meteorologist, drop us a line at *Cyclogenesis@charter.net* and well add it to the next revision and credit you for the source!

Resources:

> For immediate access to weather logs, try your local National Weather Service Office (NWS) or go online to their website: *http://www.crh.noaa.gov/* and find the Climate section. They should have a link to weather over the past year within their CWA.
>
> Daily Weather Maps: Either by subscription or using the online service, it's a way to review past weather in graphic/map format. See *http://www.hpc.ncep.noaa.gov/dwm/dwm.shtml* for information about this service.
>
> NCDC (National Climactic Data Center): A complete source on all types of weather information, records, graphs, maps, radar, satellite and much more: *http://lwf.ncdc.noaa.gov/oa/ncdc.html*

NCDC offers free access to a multitude of weather products at this address: *http://www4.ncdc.noaa.gov/cgi-win/wwcgi. dll?wwAW-MP-F*

If you prefer a more practical approach to data and not have to pay a fee every time, a CD that catalogs megabytes of weather for a single purchase. *http://ols.nndc.noaa.gov/plolstore/plsql/olstore. prodlist?category=C&subcatc=01&groupin=CDR*

The Climate Visualization system is an interactive graphing tool designed to allow visual browsing of the data available online. CLIMVIS simply requires the user to step through the data and graphing feature selection process to visually browse the data. *http://www.ncdc.noaa.gov/oa/climate/onlineprod/drought/xmgr.html*

Plymouth State Weather Center is a site where you can build Contoured 1950–2003 NCEP/NCAR Reanalysis Upper Air Maps. *http://vortex.plymouth.edu/reanal-u.html*

This product contains historical National Center for Environmental Prediction (NCEP) weather charts that are archived at the National Climatic Data Center. *http://ols.ncdc.noaa.gov/cgi-bin/nndc/buyOL-006.cgi?FNC=chart__Ancep_get_chart_htm*

For those who wish to utilize those Difax maps of the past, you can go to *http://weather.uwyo.edu/difax/index.html* and check out the selections available.

You can never have enough resource links: *http://www.refdesk.com/weath1.html*

You can link directly to an NCDC Weather Observation Station Record for a particular station (or list of stations) using identifiers *http://www.ncdc.noaa.gov/oa/climate/linktowcs.html*

Yahoo offers an area (Groups) where those with specific skills or knowledge can share it with others. It's a great way to learn more about your job or enhance your skills through others who have previous knowledge of the science. *http://groups.yahoo.com/group/forensic_meteorology/*

Pave-Cool is a computer program that shows how asphalt cools under changing weather elements. Needless to say, asphalt and the composition can often come into play in legal cases where accumulation of ice and snow is a factor. You can access this program at these two addresses; *http://groups.yahoo.com/group/forensic_meteorology/Cyclogenesis@charter.net*

There are a number of Excel spreadsheets that calculate Estimated Time of Death from various formulas and programs and can be found at those sites above. I'm sure, by the time this book is published, there will be many more posted.

Bibliography

Benjamin, D. (1993). "Forensic Pharmacology" in R. Saferstein (ed.) Forensic Science Handbook. NJ: Prentice-Hall.

Borg, W.R., and M.D. Gall, Educational Research: An Introduction, New York: Longman Inc., 1983:413–415.

BRE provides research-based consultancy, testing and certification services to customers on buildings, construction, energy, environment, fire and risk world-wide.

Borg and Gall, 413–415.

Clark, S.C., Occupational Research and Assessment, Inc., Big Rapids, Michigan, 1996.

CLP Power Wind/Wave Tunnel Facility : *Central Research Facilities* at the Hong Kong University of Science and Technology.

Combs, D., R.G. Parrish, and R.T. Ing, Death Investigation in the United States and Canada, Atlanta: U.S. Department of Health and Human Services, Public Health Service, Centers for Disease Control and Prevention, 1995.

Compu-Weather - Real-life case examples

Cyclogenesis - How to Plot and Analyze a SKEW T log p Diagram".

DPLyle,MD 2000–Timely Death

Environmental Adaptation Research Group, Atmospheric Environment Services Downsview, Ontario, Canada

Fernando Caracena NOAA, Forecast Systems Laboratory, Boulder, Colorado . Microburst study with Theodore Fujita

Feinman, Law 101: Everything You Need to Know About the American Legal System, New York: Oxford University Press, 2000. ISBN: 019 513 265 3

Haggard–2003–Definition of a Forensic Meteorologist

Hanzlick, R., "Coroner Training Needs: A Numeric and Geographic Analysis," Journal of the American Medical Association 276 (1996):1775–1778.

Jay Dix and Michael Graham - (Photo—Time of Death, Decomposition and Identification, an Atlas)

Jentzen, J.M., S.C. Clark, and M.F. Ernst, "Medicolegal Death Investigator Pre-Employment Test Development," American Journal of Forensic Medicine and Pathology 17 (1996):112-16.

J.S. Clauzel Flight Safety-DACO-1981

Karen Holmes Balestreri, Pendleton & Potocki as presented at the AMS Conference of Forensic Meteorologist. Jan/2005

Klaasen, C. (1996). "Principles of Toxicology and Treatment of Poisoning" in J. Hardman et al., Goodman and Gilman's The Pharmacological Basis of Therapeutics. NY: McGraw-Hill.

Levine, A. (1993). "Forensic Toxicology" Journal of Analytical Chemistry 65: 272–76.

Lowry, W. & J. Garriott. (1979). Forensic Toxicology: Controlled Substances and Dangerous Drugs. NY: Plenum.

MacMaster, Greg: Real-life cases examples

Moenssens, A.A.; Inbau, F.E.; Starrs, J.E. (1986). Scientific Evidence in Civil and Criminal Cases. NY: The Foundation Press.

Morten Starkeby - Cases from the archives

NTSB—Aviation Accident Investigation information referenced from: *http://www.nws.noaa.gov/om/forensic.shtml*

Paul Mason Consulting. RFID - Retail Sales

Randall B. Christison Attorney at Law as presented at the American Meteorological Society Conference of Forensic Meteorology

Saferstein, R. (1998). Criminalistics: An Introduction to Forensic Science. NJ: Prentice-Hall.

Schubert, Grillot's Introduction to Law and the Legal System, 6th Ed., Boston: Houghton Mifflin Co., 1996. ASIN: 039 574 666 3
The Ohio State University, Center on Education and Training for Employment, DACUM, 1996.

Thorntons, May 2003 -Trading Statement by

Benjamin, D. (1993). "Forensic Pharmacology" in R. Saferstein (ed.) *Forensic Science Handbook*. NJ: Prentice-Hall.

Klaasen, C. (1996). "Principles of Toxicology and Treatment of Poisoning" in J.

Hardman et al., *Goodman and Gilman's The Pharmacological Basis of Therapeutics.* NY: McGraw-Hill.

Levine, A. (1993). "Forensic Toxicology" *Journal of Analytical Chemistry* 65: 272–76.

Lowry, W. & J. Garriott. (1979). *Forensic Toxicology:*

Controlled Substances and Dangerous Drugs. NY: Plenum.

Moenssens, A.A.; Inbau, F.E.; Starrs, J.E. (1986). Scientific Evidence in Civil and Criminal Cases. NY: The Foundation Press.

Saferstein, R. (1998). *Criminalistics: An Introduction to Forensic Science.* NJ: Prentice-Hall.

Book Reviews

I have just gone through Forensic meteorology book. I can say with some authority, this is an **instant bestseller**. Please have it published immediately. Forensic community would love to read it (besides of course people in the meteorology business. Congratulations for publishing such a great work!

Dr. Anil Aggrawal is a professor of Forensic Medicine at the Maulana Azad Medical College, New Delhi

"The influence of temperature is the most important one to the object of my forensic focus: insects on corpses. This book may help as a starter not only to check for temperature but also for microclimatic changes in forensic entomology cases. It also stresses that interdisciplinary methods -- i.e., experts from different fields working together, not one expert trying to become an expert in too many fields -- that leads to truly successful understanding of what happened when and how on a scene of crime."

Dr. Mark Benecke: Expert for Collection, Examination and Scientific Interpretation of Biological Stains from Crime Scenes. Certified Forensic Biologist, International Forensic Research & Consulting, Köln Germany

"Greg has created an impressive, comprehensive, and easily understandable presentation on forensic meteorology that will be of great interest to attorneys, meteorologists, writers, and anyone who is interested in understanding the forensic field. - Designed for anyone who has an interest in how the weather plays a role in CSI."

Bob Robbins- Jenkins Group
Traverse City, Michigan

Printed in the United States
50462LVS00004B/520-591